SCIENCE FOR THE BEAUTY THERAPIST

Also by Stanley Thornes (Publishers) Ltd

Elaine Almond	*Safety in the Salon*
W E Arnould-Taylor	*The Principles and Practice of Physical Therapy*
W E Arnould-Taylor	*Aromatherapy for the Whole Person*
Ann Gallant	*Body Treatments and Dietetics for the Beauty Therapist*
Ann Gallant	*Principles and Techniques for the Electrologist*
Ann Gallant	*Principles and Techniques for the Beauty Specialist*
Ann Hagman	*The Aestheticienne*
Joyce Allsworth	*Skin Camouflage*

Beauty Guides by Ann Gallant

1 Muscle Contraction Treatment
2 Figure Treatment
3 Galvanic Treatment
4 Epilation Treatment

SCIENCE FOR THE BEAUTY THERAPIST

JOHN ROUNCE BSc

Senior lecturer in Science, Gloucestershire College of
Arts and Technology, Gloucester

STANLEY THORNES (PUBLISHERS) LTD

First published 1983 by
Stanley Thornes (Publishers) Ltd,
Old Station Drive,
Leckhampton,
CHELTENHAM GL53 0DN

Reprinted 1986 with amendments

British Library Cataloguing in Publication Data

Rounce, John F.
    Science for the beauty therapist.
    1. Beauty culture — Miscellanea
    2. Science
    I. Title
    502'.4'6467   TT957

    ISBN 0 85950 331 3

Typeset by Tech-Set, Gateshead, Tyne & Wear.
Printed and bound in Great Britain at The Bath Press, Avon.

CONTENTS

The colour spectrum

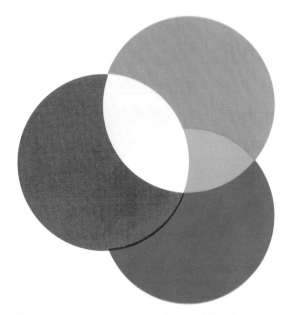

Mixing the three primary colours of light

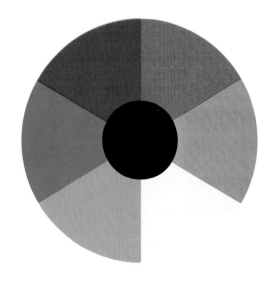

Mixing pigments — the colour disc

(a) full daylight

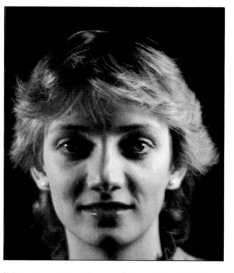

(b) a tungsten filament lamp illuminating the left side

(c) a domestic fluorescent lamp illuminating the left side

(d) another domestic fluorescent lamp illuminating the left side

The effect of light on make-up

PREFACE

Science is relevant to beauty therapy not only because it enables a therapist to understand the equipment and techniques so as to achieve the most effective treatment but also to help when the client asks how her treatment works. In addition, the competent therapist, especially the one in charge of a clinic, will want to understand the workings of water supplies, electricity and other services as well as the lighting and other environmental features of the clinic. Equally important is the greater satisfaction for those therapists who understand the scientific principles upon which their work is based. There is too a need to read and understand literature that will describe new techniques and advertisers' brochures which are increasingly giving scientific detail.

It is with these considerations in mind that this book has been written. However, the student whose ambitions extend, at present, no farther than passing a beauty therapy examination has not been forgotten.

Students are advised to read the text from the beginning. Footnotes are added for those requiring fuller explanations, and there is a summary of the main points at the end of each chapter.

The topics discussed cover a wide range but some aspects of anatomy have been omitted because these are described in books on therapy techniques. Apart from this, the text should cover most of the therapist's needs and suit the requirements of examinations such as the City and Guilds of London Institute examination (Course 761). The exercises at the end of each chapter include some questions from examinations of the CGLI and I am grateful for permission to reproduce these.

John Rounce
Gloucester 1983

A NOTE ON SAFETY

Throughout this book an attempt has been made to emphasise the need for safety in the salon, particularly in regard to electricity and chemical products. The advice given has been checked and is believed to be sound. However, the author and publishers will not be held responsible in the event of any accident involving a reader of the book, whether as the result of wrong information or a misunderstanding on the part of the reader.

1

Measurements

'36-26-36, and weight 8 stone 7 pounds'. *Measurements* such as these need no further comment, but female figures are not always as good as that in Fig. 1.1. So we begin.

Science has been described as the study of *matter* and *energy*, and it has always been very much concerned with measuring. Our opening quotation gives one illustration of the importance of measurements in beauty therapy. The tissues of the body and the materials used in the beauty salon are the kinds of matter that concern us here, and you will find that many beauty treatments entail putting energy into the body. Dieting, however, is concerned with limiting the energy associated with food intake.

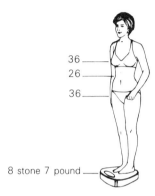

Fig. 1.1   Measurements are important to the beauty therapist

---

UNITS OF MEASUREMENT

SI units

In our introduction the figures 36-26-36 described bust, waist and hip measurements. These sizes are clearly in inches, meaning that the *unit* used is the *inch*. Instead we can write 92-66-92 using the *centimetre* as the unit of measurement. Similarly 8 stone 7 pound (119 lb) can be replaced by 54 kilogram. The kilogram is the unit recommended by the international system of units (SI) employed in this book.*

⊒ MEASUREMENT OF LENGTHS, HEIGHTS AND OTHER DISTANCES ⊒

The SI unit here is the *metre* and the abbreviation for metre is m (*not* capital M). Some useful multiples and subdivisions of the metre are listed overleaf (see also Tables 1 and 2, p. 291):

---

*Scientists and engineers are already working almost entirely with metres and subdivisions (e.g. centimetres) and multiples (e.g. kilometres) of the metre. This practice is preferred to the use of imperial units such as the inch, foot, etc. and is one of the recommendations of the International System of Units and Measurements (Système International or SI system). Similarly this system employs the kilogram in preference to the pound. In fact the SI system, which has been accepted by a large number of countries, names preferred units for most measurements. It also specifies the subdivisions and multiples which are preferred and the abbreviations that are acceptable.

kilometre — abbreviation km — meaning one thousand metres

centimetre — abbreviation cm — one hundredth of a metre or 1/100 m

millimetre — abbreviation mm — one thousandth of a metre, 1/1000 m

micrometre — abbreviation μm — one millionth of a metre, 1/1 000 000 m

nanometre — abbreviation nm — one thousand-millionth of a metre, 1/1 000 000 000 m

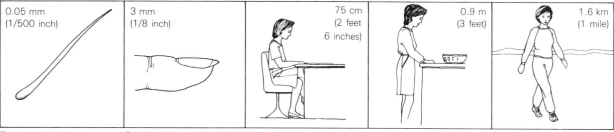

| 0.05 mm (1/500 inch) | 3 mm (1/8 inch) | 75 cm (2 feet 6 inches) | 0.9 m (3 feet) | 1.6 km (1 mile) |

Thickness of human hair | Growth of finger nail in one month | Suitable height for a desk | Suitable height for a bench | Distance walked to use 300 Calories

Fig. 1.2   Examples of measurements

AREA, VOLUME AND MASS

Area

Area means 'amount of surface'. A square surface measuring 1 metre by 1 metre is said to have an area of 1 *square metre* or 1 *metre squared* (1 m²). For any rectangular or square surface the area can be calculated from the formula

Area = Length × Width

An example is shown in Fig. 1.3.

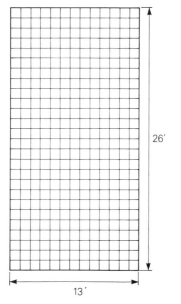

Carpeting a rectangular floor measuring 4 m (13 feet) by 8 m (26 feet)

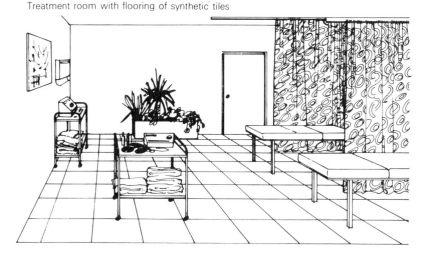

Treatment room with flooring of synthetic tiles

Area = 4 m × 8 m = 32 m² or 13 × 26 = 338 ft²
(Using 1 foot square carpet tiles would require 338 tiles)

Fig. 1.3   The area of a clinic floor

Imperial to metric conversions

1 foot is very nearly equal to 300 mm, so that

Number of mm = Number of feet × 300*

Volume

The *volume* of something is the 'size of the space it occupies'.

Volume can be measured, for example, in teaspoonfuls, cupfuls, pints, litres, fluid ounces or cubic centimetres, but the SI unit is the cubic metre.[†]

The volume of a rectangular object can be calculated from the formula

Volume = Length × Width × Height

Note that a litre is a volume of 1000 cm³ or 1/1000th of a metre³. Also, one meets the abbreviations cc (cubic centimetre) and ml (millilitre); each of these means cm³. A centilitre (cl) is 1/100 litre or 10 cm³.

The fluid ounce is a unit of volume, being the volume occupied by one ounce of water.

Some methods of measuring volumes are shown in Fig. 1.4.

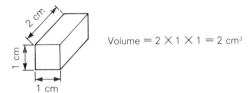

(a) Volume = 2 × 1 × 1 = 2 cm³

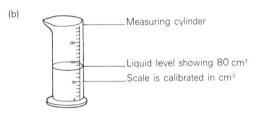

(b) Measuring cylinder
Liquid level showing 80 cm³
Scale is calibrated in cm³

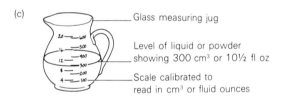

(c) Glass measuring jug
Level of liquid or powder showing 300 cm³ or 10½ fl oz
Scale calibrated to read in cm³ or fluid ounces

Fig. 1.4   Some methods for measuring volume

Mass and weight

The *mass* of a thing (we usually say 'of an object') means the amount of *matter* which it contains (Matter is the word used to describe what everything around us is made of. Water, air, metals, wood, cloth, plastics, creams, oils, waxes and flesh are, of course, examples of matter.) Common units for mass are the pound (lb), gram (g) and kilogram (kg).

All objects attract each other simply because of their masses. This phenomenon is called *gravity* or *gravitation*. These attractions are weak and not usually noticed except when an object is attracted by a very massive thing like the earth. The force of gravity on an object is called the *weight* of the object.

Most people possibly including beauty therapists will say 'weight' when they mean mass. This custom should not cause any serious problem in beauty therapy work.

---

*It also follows that 1 yard (= 3 feet) is equal to 3 × 300 mm. As an example of converting inches to mm consider a length of 40 inches. The corresponding number of mm is $\frac{40}{12} \times 300$ mm = 1000 mm or 1 m.

[†] A cubic metre of (say) a liquid would just fill a space measuring 1 metre by 1 metre by 1 metre and is often described as being a volume of 1 metre cubed or 1 m³ (the superscript 3 denoting that three lots of metres have been multiplied together).

Mixing ingredients

Suppose a mixture is made using one measured volume of one ingredient added to 9 equal volumes of another. The mixture can then be described as a 10% volume for volume mixture, i.e. 10% v/v. ('Per cent' means 'per 100 parts', of course, so that 10% means 10 parts of one ingredient for every 100 parts of the mixture.)

Another way of specifying the relative proportions of the ingredients in a mixture is the weight for weight percentage (% w/w). For example, 3 g of powder mixed with 9 g of oil produces 12 g of mixture of which 3 g or ¼ (25%) of the mixture by weight is powder. The powder concentration is 25% w/w.

$$\text{Concentration} = \frac{\text{Volume of ingredient}}{\text{Volume of mixture}} \times 100\% \text{ v/v}$$

or

$$= \frac{\text{Mass of ingredient}}{\text{Mass of mixture}} \times 100\% \text{ w/w}$$

When a solid is dissolved in a liquid (solutions are discussed in Chapter 9) the concentration is often described by the number of grams dissolved per litre of liquid.

Fig. 1.5 Weights and measures are frequently shown on bottle labels

FORCE, WORK AND ENERGY

Force

A *force* is a push or a pull.

A force is essential to speed up or slow down an object.

If a force is applied to an object it will begin to move or it will change its speed unless another equal force is also applied and is in exactly the opposite direction.

A force can be measured in pounds-force or kilograms-force but the SI unit is the *newton* (N).

A kilogram force (kgf) is a force equal to the force due to gravity on a 1 kg object.

1 kgf = 10 N approximately

Some examples of forces are shown in Fig. 1.6.

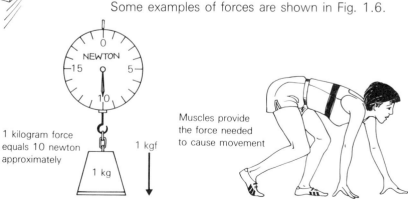

The force due to gravity on an object moves it downwards. The force due to gravity on 1 kg is called a kilogram force

1 kilogram force equals 10 newton approximately

1 kgf

Muscles provide the force needed to cause movement

Fig. 1.6 Examples of forces

Reaction forces

Often a force is applied to an object, and as a result the object starts to move but then the movement stops, sometimes after only a microscopic movement has occurred. The movement itself has produced an opposing force, called a *reaction force,* which has stopped the movement. For example, during massage each push upon the body moves the body tissue so that it is squeezed and pushes back. Some examples of reaction force are shown in Fig. 1.7.

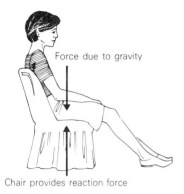

Force due to gravity

Chair provides reaction force

(b) Resting in a chair

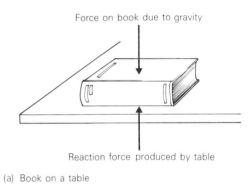

Force on book due to gravity

Reaction force produced by table

(a) Book on a table

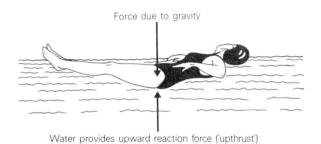

Force due to gravity

Water provides upward reaction force ('upthrust')

(c) Floating

Fig. 1.7 Examples of reaction force

Work

Here we confine our attention to physical work, for example, lifting a heavy object, pushing a car or climbing a hill, rather than mental work (see Fig. 1.8).*

5kg

Lifting

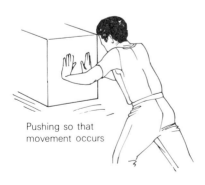

Pushing so that movement occurs

Fig. 1.8 Some examples of work

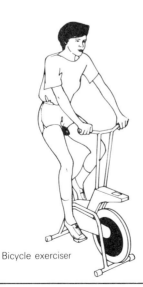

Bicycle exerciser

*We all agree that physical work is something that makes us feel hot and tired. In addition, we should note that a push or a pull, that is to say a force, is always being used when work is done. For example, a car is being pushed from behind, and the work is greater if a large push is needed or if the pushing continues for a long distance along the road.

The amount of work done is measured by the product of the force and the distance moved (in the direction of the force, of course):

Work done = Force × Distance moved

Using SI units, namely newton for force and metre for distance, the answer obtained for the work is in units called *joules* (abbreviation J).

Work done by lifting

To lift a 5 kg object slowly, we must use an upward force slightly greater than the 5 kgf (or 50 N approximately) due to gravity. The work involved (in joules) in this operation is, therefore, approximately 50 times the number of metres through which the object is raised. For example, 5 kg raised through 2 m means 100 J of work is done.

A 50 kg person climbing 4 m upstairs has done 2000 J of work (Fig. 1.9).

Work done by lifting in J = Mass in kg × 10
× Height in m

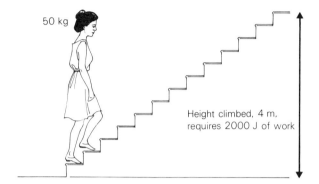
50 kg

Height climbed, 4 m, requires 2000 J of work

Fig. 1.9 Work done when climbing stairs

Work done when using an exercise bicycle

For a bicycle with the attachments as shown in Fig. 1.10 the work done can be calculated as follows. The forces pulling on the two ends of the brake band are read from the spring balances (in newtons) while the bicycle is being driven at a steady speed. These two figures are subtracted and the answer obtained is multiplied by the perimeter of the wheel (distance around the rim) and then by the number of revolutions of the wheel that are completed. The answer is in joules. This can be used later to obtain some idea of the number of Calories used during the exercise.

Some exercise bicycles are fitted with a meter that gives the answer for the braking force multiplied by the wheel radius (called the *moment* or *torque*). The work done is then given by:

Work = Moment × 2 × π × No. of revolutions

where $\pi = 3.14$ or $\dfrac{22}{7}$.

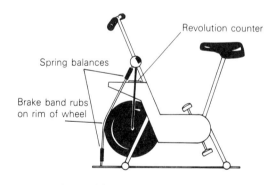
Revolution counter

Spring balances

Brake band rubs on rim of wheel

Fig. 1.10 An exercise bicycle

Static work

When a force is produced by muscle action it is possible for work to be done even when there is no visible movement. This happens, for example, if a person pushes against an immovable object such as a brick wall. No movement is seen and the muscle length does not change, i.e. the muscle is being used *isometrically*.*

Energy

Whenever work is to be done *energy* has to be used. Energy is 'the ability to do work'. A person's energy is obtained from the food that he or she eats (see Fig. 1.11).

Energy is needed for work to be done rather like money has to be used to buy goods. Petrol (or gasoline) or similar fuel is used to provide the energy needed to move many vehicles. Food is eaten to enable a person to do the day's work. The food or fuel contains the energy needed.

How much energy contained by something is best described by specifying the number of joules of work which can be obtained from the energy. Hence the amount of energy that allows 1 joule of work to be done is itself described as 1 joule of energy. One slice of buttered toast, for example, contains about 600 000 J (600 kJ) of energy.

The SI unit for energy and for work is the joule (J).

1 calorie equals 4.2 J and 1 Calorie (capital C) equals 1 kilocalorie or 4200 J.

140 Calories
(590 000 joules)

One slice of buttered toast

20 Calories
(84 000 joules)

One cup of tea with milk (but no sugar)

600 Calories
(2 500 000 joules)

Fish and chips

Fig. 1.11   The energy content of some items of food

Forms of energy

Just as money can be in the form of notes, coins or cheques so energy is found in many forms.

Energy possessed by foods and fuels is called *chemical energy*.

A moving object possesses another form of energy called *kinetic energy*.†

An object above the ground has energy called *gravitational potential energy*. The ability of this object to do work becomes evident if it is allowed to fall because it will then acquire kinetic energy which we know can produce work.

Energy is described as potential energy when it is not due to movement but to the position of the object or the relative positions of its parts. In the example just described, the energy is due to the object's separation from the earth which attracts it.

Another form of potential energy is the *elastic* or *spring energy* which is contained, for example, in a stretched spring or a compressed spring.

So far we have two main types of energy: kinetic energy and potential energy such as gravitational, elastic or (see Chapters 3 and 18) chemical energy. *Electromagnetic radiation*, including light and radiant heat, is a third form of energy, and this is discussed in Chapters 15 and 17.

---

*Presumably work is being done because all the muscle fibres are busily tugging and relaxing, contracting and lengthening, over and over again. Thus there is, in fact, a lot of microscopic movement in the muscle.
This is quite different from leaning against a wall when no work is being done at all.

†That a moving object possesses the ability to do work is obvious from the fact that if it were to hit another object, that object could be made to move, like a car being pushed, so that work is done.

Conservation of energy

This is the name given to the rule that when energy is used to do work it becomes a different form of energy but there is as much energy afterwards as there was before. In other words,

Energy cannot be created or destroyed. *

To illustrate the conservation of energy we can consider that some chemical potential energy obtained from food is used when a person throws a tennis ball as shown in Fig. 1.12.

The energy is changed into kinetic energy. This subsequently becomes gravitational potential energy, then kinetic energy, and lastly, it is seen as the energy of the ripples on the water. The energy concerned in this sequence of events could be about 6 joules. At each change from one form of energy to another this number of joules of work is done so that, for example, when the ball is thrown the force of the hand upon the ball does this much work. †

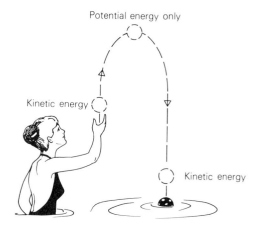

Fig. 1.12　The conservation of energy

Heat

Heat is energy, as explained later in Chapters 7 and 10.

Consider some water that is being warmed up. Heat energy is being put into it and, as a result, its temperature rises. This is what we mean by *heating*. The heat energy has become energy of vibration and other microscopic movements of all the little bits (*molecules*) of which the water is made. Every molecule is shaking about (*vibrating*) and the warmer we make the water the more its molecules vibrate.

Power

This term is used to describe the rate at which work is being done, in other words, the rate at which energy is being converted from one form into another:

Power = Work per second or Energy per second

A power of 1 joule per second is called a *watt* (W).

Efficiency

The efficiency of an energy-converting process is given by the formula:

$$\text{Efficiency} = \frac{\text{Energy obtained in the required form}}{\text{Total energy converted}} = \frac{\text{Useful work done}}{\text{Total work done}}$$

*This situation is similar to spending money in buying goods. As much money exists after the purchase as there was before, but a change has occurred. It is now money in someone else's pocket. In the case of energy the change is from one form of energy to a different form. Again it must be emphasised that work cannot be done without energy just as goods cannot generally be obtained without money.

† If energy is never destroyed how can the world become short of energy? The answer is that it can become short of energy in a form that can be used. There is a tendency for energy to become spread out so that it is difficult to use.
In the example of the tennis ball falling into water it would clearly be difficult to change this energy back into kinetic energy of the tennis ball.

This formula gives an answer for the efficiency which is a fraction. This can be multiplied by 100 to obtain the answer as a percentage.

For example, in the conversion of the chemical energy in muscles to the kinetic energy of limb movement only about one-fifth of the potential energy is converted to kinetic energy, while the other four-fifths heats the body.

$$\therefore \textbf{Efficiency} = \frac{\textbf{Useful work done}}{\textbf{Total work done}} = \frac{1}{5} = \frac{1}{5} \times \textbf{100\%}$$

$$= \textbf{20\%}$$

Efficiency really tells us how successful the energy change is.

SUMMARY

- Area = length × width and is measured in square feet or square metres (metres squared, m^2).

- Volume = length × width × height and is measured in cubic feet or cubic metres or metres cubed (m^3).

- The mass of an object tells us how much matter it contains or how much it weighs.

- Concentration, v/v = $\frac{\text{Volume of ingredient}}{\text{Volume of mixture}} \times 100\%$

- Concentration, w/w = $\frac{\text{Weight of ingredient}}{\text{Weight of mixture}} \times 100\%$

- Concentration, as weight per unit of volume is measured in g/cm^3 or g per litre, etc.

- A force is a push or a pull and is measured in kg force (kgf) or in newtons (N).

- Work = force × distance. The unit for work is the joule (J).

- Work done in lifting can be calculated from mass in kg × 10 × height in m. The answer is in joules.

- Energy is the ability to do work. It is measured in joules or in calories.

- 1 kilocalorie is 1000 calories and can be written as 1 Calorie (capital C). A slice of buttered toast provides approximately 140 Calories.

- Energy exists in many forms. It may be: potential energy (chemical, gravitational, for example), or light (or other electromagnetic radiation), or kinetic energy.

- When work is done, energy is changed from one form into another form and energy is conserved, i.e. we have as much energy after as we had before.

- Energy used (for example, energy obtained from food) does not all become changed into the required form (for example, kinetic energy of body movement) because there is a tendency for some heating to occur.

- Efficiency tells us the proportion of energy which is changed into the required form, i.e.

$$\textbf{Efficiency} = \frac{\textbf{Useful work}}{\textbf{Energy used}}$$

- Power is work per second. 1 joule per second is called a watt (W).

EXERCISE 1

(Please refer to Tables 1 and 2 on p. 291 as necessary.)

1. A certain wax heater has a capacity of (i.e. it can hold) 11 pounds of paraffin wax. How much is this in kilograms? A smaller model of this heater has a capacity of 2 kilograms. Express this in pounds.

2. A convenient height for the surface of a bench at which a person is to stand is 900 mm. What is this height: (a) in centimetres, (b) in metres, (c) in feet approximately?

3. A certain needle used for epilation (hair removal) has a diameter of 0.005 inches and a length of one inch. What is the value in millimetres of: (a) its radius, (b) its length?

4. A plate electrode consisting of a rectangular sheet of metal has dimensions of 50 mm by 40 mm. Calculate the area of one face of this electrode: (a) in mm^2, (b) in m^2.

5. What area of carpet is needed for a room which is rectangular and measures 7 m by 4 m?

6. Convert $40'' - 30'' - 41''$ to centimetres.

7. What is the full name of each of the units used in the following measurements?

(a) 3″ (b) 3 kilos (c) 3 lb (d) 3 oz

8. How many minutes equal: (a) 1¾ hour, (b) 0.1 hour, (c) 1/3 hour?

9. Calculate the number of teaspoonfuls equal to $12 \ cm^3$.

10. Calculate the number of litres corresponding to 2 gallons.

11. The water-boiling tank of a certain steriliser measures 30 cm by 15 cm by 20 cm depth and is to be ¾ filled with water. Calculate the volume of water needed: (a) in cm^3, (b) in litres.

12. How many treatments will be possible from a half-litre bottle of lotion if $1 \ cm^3$ is needed per treatment?

13. If the human body normally contains 15% fat by weight, how much fat is carried by a normal person weighing 50 kg?

14. A hair-pull composition is to be made from rosin 69% (by weight), beeswax 20%, together with other ingredients amounting to 11%. How many grams of beeswax will be needed to make 1 kg of mixture?

15. A saline solution for iontophoresis (see Chapter 5) is being made up by dissolving 25 oz of salt (sodium chloride) in 900 g of water. Calculate the strength of the solution as a weight for weight percentage (see Table 2).

16. One type of after-shave lotion contains mycia oil and other ingredients dissolved in water. For 500 ml of solution 4 ml of oil is used. What is the concentration of the oil by volume?

17. A perfume can be made by dissolving lilac floral fragrance in phenyl ethyl alcohol. How many grams of alcohol must be added to 6 g of fragrance to produce a 30% w/w solution?

18. Answer each of the following questions using the term *reaction force*.

(a) Why, if a hair is pulled gently, does it move a little and then stop?

(b) If a person squeezes hard on a grip exerciser dumbbell, the skin of the hand may be depressed and look pale afterwards. Why?

19. (a) What is the size of the downward force due to gravity on a 6 kg object? Give the answer: (a) in kgf, (b) in N.

(b) When this object is held in the hands, what force must the hands provide to stop the object falling?

(c) Approximately what force is needed to very slowly raise this object?

(d) Approximately how much work is done when lifting the object through a height of 1½ metres?

20. If it were found that 100 J of energy were used to provide 70 J of useful work, (a) how much energy is wasted, (b) what is the efficiency of the energy-conversion process?

21. When an average person is sitting, energy is being used at a rate of 100 watts because of the essential processes taking place in the body even in the absence of exertion of any kind.

How many joules of energy will be used in this way in 24 hours?

22. How much work is being done per minute to operate an exercise bicycle at a speed of 60 revolutions per minute if a crank torque meter on the bicycle indicates 10 newton metres?

23. If the work done per minute when using an exercise bicycle were 3600 J, calculate the power being used assuming that the cycling is being done steadily.

Also calculate an approximate value for the energy being used by the cyclist per second if the efficiency of her cycling is 20%.

2

PRESSURE AND FLUIDS

PRESSURE AND LIQUID FLOW

Pressure

If a shoe has a slim heel so that it has only a small area in contact with the floor, it may damage the floor when the weight of the person standing in the shoe is pressing downwards. A carpet may be dented for the same reason. If the person's weight is spread over a larger area the floor is not likely to be harmed.

Pressure means force divided by the area over which it is spread.

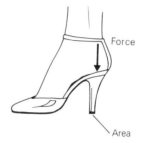

$$\text{Pressure} = \frac{\text{Force}}{\text{Area}}$$

It is measured usually in newton per metre^2 (also called pascal, Pa).

In the case of the shoe pressing on the floor the damage is likely to occur when the pressure is high. Similarly, during manual massage, the effect upon the body tissue is decided by the pressure applied by the hand because not only is the force used important but also the area of contact between the therapist's hands and the client's body.

Pressure tells us about the 'squeeze' on the floor or on the flesh (Fig. 2.1).

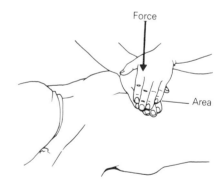

Fig. 2.1   Pressure is the force divided by the area

Fluids

Gases and liquids are called *fluids* because they can flow.

The pressure under a liquid

Anything with liquid above it is squeezed by the weight of the liquid pushing down on it. The pressure on a submerged surface can be studied experimentally as suggested in Fig. 2.2 (overleaf).

11

In this experiment, the deeper the tube is held below the liquid surface the more the liquid pushes the surface of the air into the tube (i.e. squeezes the air more). The greater depth gives a greater pressure. Changing the shape of the container does not affect the pressure produced. Only depth affects it.

Pressure increases with depth.

This explains why varicose veins often occur low down in the legs (Fig. 2.3).

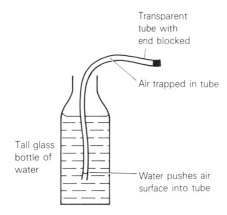

Fig. 2.2 To show that pressure increases with depth

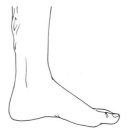

Fig. 2.3 Varicose veins are more likely to occur low down in the legs where the blood pressure in the veins is greatest

Atmospheric pressure

The air around us, which is the earth's atmosphere, extends to thousands of metres above ground level. Consequently the surface of a person's head at ground level experiences a very considerable pressure, in fact a pressure of about 1 kgf on each square centimetre, due to the air above.

Atmospheric pressure is about 1 kgf per cm^2, or 10 N per cm^2. Its value is affected by the weather, and a meter for measuring atmospheric pressure (a *barometer*) can be used as a guide to the weather expected.

As shown in Fig. 2.4, normal atmospheric pressure can be described as 760 mm of mercury pressure.

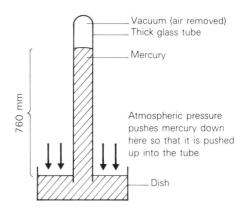

Fig. 2.4 A simple mercury tube barometer

The pressure in a fluid acts in all directions

If a pressure is applied on some fluid, this fluid will push out in all directions.

A simple example of this is illustrated in Fig. 2.5 where a downward pressure on a tube of toothpaste causes paste to push out in all directions. Note that atmospheric pressure presses upon our bodies, not only downwards upon our heads and shoulders, but inwards on every part of the body surface because the surrounding air pushes in all directions. Also the pressure within a person's mouth and lungs is not much different from the atmospheric pressure. If this were not so, we could expect the body walls to be pushed inwards by the inward atmospheric pressure.

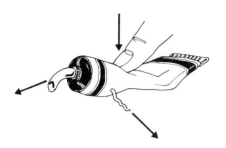

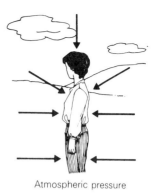

Atmospheric pressure

Fig. 2.5 Fluid pressure acts equally in all directions

Flow caused by pressure differences

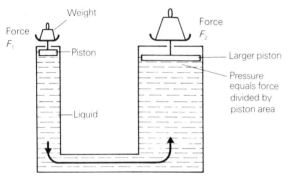

Fig. 2.6 Liquid flow

In Fig. 2.6 it might be thought that to make the liquid flow in the direction shown the force F_1 would simply have to be bigger than force F_2. In fact, flow occurs and the right-hand piston moves up (the left-hand one moving down as well, of course) as soon as the pressure on the left-hand side exceeds that on the right. Because of the small area of the left-hand piston this occurs when F_1 is quite small.

Fluids move when there is a pressure difference.

Flow occurs from a high-pressure place to a low-pressure place.

A liquid can also fall because of gravity, such as running down a pipe. An opposing pressure difference may stop a liquid's fall.

COLD WATER SUPPLY

A domestic cold water system

A typical cold water supply system to a house is illustrated in Fig. 2.7 (overleaf).

Without the cold storage tank a problem would arise when the hand-basin, the bath and WC all require water at the same time. The main supply pipe would be unable to provide water fast enough.

Drinking water is preferably taken from the main supply; dirt may get into water in the cold tank.

The ball valve allows entry of water as soon as the water level in the tank drops. Once the level of water has been restored it will have lifted the floating ball so that a pad

(*washer*) is pushed against the end of the supply pipe to stop further entry of water.

Except when the demand is great, the pressure in the main supply pipe is high while that in the tank-supplied piping is lower and is decided by the height of water (i.e. the head of water) above the point concerned. If a new piece of equipment is to be supplied, such as an extra shower, then it is important to ensure that there is sufficient head of water to operate it.

In the event of leaks, use may be made of the *stopcocks* to prevent further entry of water and the cold tank can be emptied by opening one or more of the cold water taps.

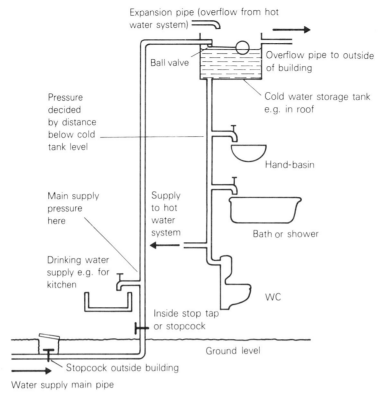

Fig. 2.7 A domestic cold water system

═ BLOOD AND LYMPH CIRCULATION ═

─ Blood circulation ─

Fig. 2.8 is a simplified diagram of the human blood circulation system.

When the right ventricle contracts and squeezes on the blood inside it, the increase of pressure makes the blood flow out through the pulmonary arteries to the lungs. Here the blood receives oxygen and carbon dioxide is removed from it. The oxygenated blood returns to the left auricle of the heart via the pulmonary veins.

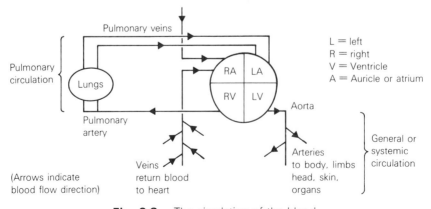

Fig. 2.8 The circulation of the blood

Next, the contraction of the left auricle moves the blood into the left ventricle which in turn pumps the blood into the aorta artery. From here the blood flows through arteries, wide ones at first which then divide into narrower ones, and finally enters very narrow capillaries in all the various tissues of the body.

Clear liquid, containing oxygen, nutrients, other chemicals and white corpuscles (Chapter 20), passes from the blood in the capillaries into the surrounding tissues. This fluid is called *lymph*. Carbon dioxide from the tissues enters the blood.

Return of the blood to the heart is through veins. For this return journey the pressure difference needed to cause the blood to flow arises largely from the movement of nearby muscles. As leg muscles, for example, are used, they squeeze the veins in the legs. Valves within the veins ensure that the resulting blood flow can only be in the direction towards the heart. Thus the blood returns to the right auricle ready to circulate again through the lungs and then through the rest of the body again.

Clearly the pumping of the auricles and ventricles must be properly timed. *Ventricular fibrillation* consists of contractions of the heart which are not suitably timed and fail to adequately circulate the blood. This is the cause of death in many cases of electric shock.

The normal pulse rate (and therefore heart beat rate) is about 70 per minute. This figure is increased by exercise and by heat treatments.

Systole is the name given to contraction of the heart. Auricular systole follows ventricular systole and then all chambers are at rest (*diastole*). It is this sequence, called the *cardiac cycle*, which repeats about 70 times per minute. The maximum blood pressure, caused by systole, should be about 100 to 150 mm of mercury depending upon age, with diastole being something like 70 mm.

The lymphatic system

In addition to the system of blood-carrying vessels, which has just been described, there is the *lymphatic system*. This system's function is to take lymph from the tissues with waste material that it now contains and transport it to the subclavian region, near the neck, where it is returned to the blood circulation system.

Pumping in the lymphatic vessels is again due to movement of nearby muscles. The larger lymphatic vessels have, like veins; valves to ensure that the lymph flows only in the required direction.

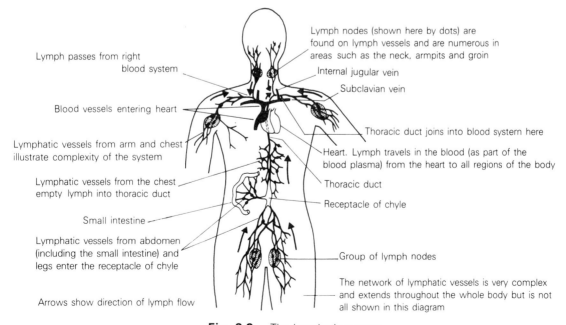

Lymph nodes (shown here by dots) are found on lymph vessels and are numerous in areas such as the neck, armpits and groin

Lymph passes from right blood system

Internal jugular vein

Subclavian vein

Blood vessels entering heart

Thoracic duct joins into blood system here

Lymphatic vessels from arm and chest illustrate complexity of the system

Heart. Lymph travels in the blood (as part of the blood plasma) from the heart to all regions of the body

Lymphatic vessels from the chest empty lymph into thoracic duct

Thoracic duct

Receptacle of chyle

Small intestine

Lymphatic vessels from abdomen (including the small intestine) and legs enter the receptacle of chyle

Group of lymph nodes

The network of lymphatic vessels is very complex and extends throughout the whole body but is not all shown in this diagram

Arrows show direction of lymph flow

Fig. 2.9 The lymphatic system

SUMMARY

- Pressure is force divided by area and describes 'squeeze'.

- The pressure in a liquid or gas increases with depth and is the same for all directions.

- Atmospheric pressure is about 76 cm of mercury. It is measured with a barometer.

- A pressure difference can make liquid or gas flow or prevent its fall.

- Cold water is pumped through main pipes. A cold water tank allows for sudden high demands from baths, sinks, etc.

- Drinking water is taken from the main supply, not from the tank.

- In the event of leaks, use is made of the stopcocks. There is usually one inside the building and one just outside. Taps may be opened to empty the cold water tank.

- Blood circulation in the body is largely due to the heart acting as a pump but is assisted by movements of muscles adjacent to veins.

- The flow of lymph is also caused by movement of muscles near to lymphatic vessels but some lymphatic vessels have muscles in their walls.

- Blood leaves the heart from the right ventricle, passes through the lungs and returns to the left auricle. It then leaves the left ventricle, passes round the body and returns to the right auricle.

EXERCISE 2

1. Calculate the pressure in kilogram force per square metre beneath the shoe of a person weighing 10 stone 3 pound (65 kg) standing on one foot, given that the area of the shoe in contact with the ground is 130 cm^2.

2. What difference is there in meaning between N per m^2, N/m^2, N m^{-2} and pascal (Pa)?

3. Which of the following statements is incorrect? The pressure in a fluid:

(a) can be measured in cm of mercury

(b) acts in all directions

(c) is greatest higher up in the fluid

(d) is important regarding liquid flow.

4. What is a piece of equipment for measuring atmospheric pressure called?

5. Liquid only flows in a horizontal pipe if there is a ___.

6. Suggest what *immediate* action should be taken in each of the following situations:

(a) a serious leak occurs from a kitchen tap on a main supply pipe

(b) the cold water tank is overflowing

(c) the hand-basin cold tap will not turn off.

(Note that, if the main supply is shut off and then hot water is used, the hot water system can run short of water. Therefore, to be safe, the hot water heater should be switched off whenever a stop tap is closed.)

7. Which of the following statements is/are correct?

(a) Arteries carry blood away from the heart

(b) Most arteries carry oxygenated blood

(c) Valves are found in many veins

(d) Oxygenated blood enters the left auricle of the heart.

8. Name a part of the body where lymph nodes are numerous.

9. A hydraulic lifting mechanism is used on some clinic chairs. The fluid used is oil (although the term 'hydraulic' might suggest water). If a 60 kg client is to be raised slowly or supported by a piston of area 10 cm^2:

(a) what pressure must be produced in the oil?

(b) what force must be applied on the smaller piston of area 1 cm^2 to produce the required pressure in the oil?

10. Name the causes of the pressure differences which produce the flow in each of the following cases:

(a) arterial blood

(b) venous blood

(c) lymph flow in lymphatic vessels

(d) water in a pipe supplied by a cold water tank.

3

THE STRUCTURE OF MATTER

MATTER

The building bricks of matter

As explained in Chapter 1, *matter* is what things are made of. Matter occurs as *solids, liquids* and *gases* (and gases include *vapours*). All of these so-called *states* are familiar to the beauty therapist (Fig. 3.1), and it is quite often easy to change matter from one of these states to another.

All kinds of matter are made from *protons, electrons* and *neutrons*.*

The differences between the various kinds of matter that we see around us are due to the different ways in which protons, electrons and neutrons can be grouped together. In the same way, different houses, castles and hotels could be built using no more than three kinds of bricks.

Zinc oxide, titanium dioxide and talc are white powders used in foundation creams, face powders and talcum powders

Castor oil is a liquid used in lipsticks

The smell from a perfume is due to invisible vapours entering the nose

Fig. 3.1 Solid, liquid and vapour illustrated by some cosmetic materials

Atoms

Protons, neutrons and electrons are usually arranged in small groups called *atoms*.

The central part of the atom is called the *nucleus* and this is where the protons and neutrons are found. Electrons are found outside the nucleus and they move round it unceasingly in paths called *orbits*.† Examples are shown in Fig. 3.2 (overleaf).

---

*Protons are very small things and certainly far too small to be seen even with any kind of microscope. All protons are exactly alike. Electrons are all exactly the same as each other but they are even smaller than protons. Neutrons are about the same size as protons. Electrons always attract protons, this pull being strong when an electron and a proton are close together. This is discussed in Chapter 4.

† Although the electrons move around the nucleus at high speed, they do not fly off because they are attracted towards the protons in the nucleus, just enough to stop their escape.

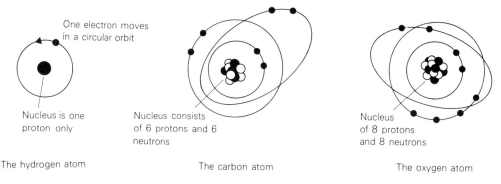

One electron moves in a circular orbit

Nucleus is one proton only

The hydrogen atom

Nucleus consists of 6 protons and 6 neutrons

The carbon atom

Nucleus of 8 protons and 8 neutrons

The oxygen atom

Fig. 3.2 The construction of some atoms*

The names of atoms. A list of names of atoms and recognised abbreviations is given in Table 3 on p. 292.

Molecules

Atoms attract each other for various reasons, and sometimes the attraction is strong enough for them to group together to form *molecules*. For example, hydrogen atoms are not usually found alone. They group together in pairs or, in other words, form two-atom (*diatomic*) molecules. Similarly ordinary oxygen molecules are diatomic while the molecule formed of three oxygen atoms is *ozone* which has some germicidal properties useful to the beauty therapist. It is therefore a *triatomic* molecule.

The chemical symbols for hydrogen and oxygen are H and O. The molecules are denoted by H_2 and O_2 so as to indicate the number of atoms in the molecules. Ozone is denoted by O_3. The structure of the hydrogen molecule is shown in Fig. 3.3.

The holding together of atoms to make molecules is called *bonding.*[†]

When bonding is due to the sharing of electrons it is called *covalent* bonding.

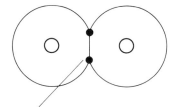

The two electrons are shared and the two atoms hold together

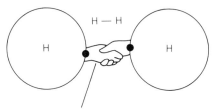

There is a bond between the two atoms (due to the sharing of two electrons) almost as if they have an arm each to link together

Fig. 3.3 The hydrogen molecule (H_2)

---

*It is *carbon* that gives the black colour to soot, and it is used for obtaining the black colour of mascara. Even the smallest visible speck of carbon black is made up of vast numbers of carbon atoms each constructed as shown in Fig. 3.2. Iron, which is commonly made into steel for clinic furniture, for instrument cases and electrodes, has atoms made up of 26 protons, 30 neutrons and 26 electrons.
Note that the number of electrons in an atom normally equals the number of protons.
Mercury, which is so often seen in thermometers, is a metal but it is a liquid at ordinary temperatures. It is quite heavy stuff because each atom contains as many as 80 protons, 121 neutrons and, of course, 80 electrons. It has the typical shiny appearance of a metal and allows electric current to flow through it if required.

[†] Why do atoms attract each other and so form molecules? Well, let's consider the example of the H_2 molecule. Although hydrogen atoms usually 'want' most of all to have the same number of electrons as protons, they 'like' to have two electrons in orbit. By forming H_2 molecules each atom has a share in two electrons, yet there are still two electrons for two protons. This is illustrated in Fig. 3.3

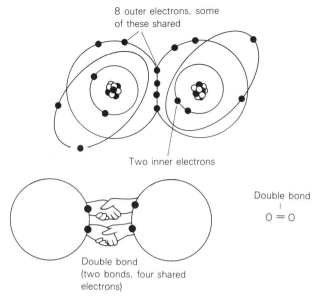

8 outer electrons, some of these shared

Two inner electrons

Double bond (two bonds, four shared electrons)

Fig. 3.4 The O_2 molecule is formed by the sharing of four electrons, i.e. by formation of a double bond

Double bond

$$O = O$$

With most atoms other than hydrogen, the reason for their joining by covalent bonding is that the atoms 'like' to have eight outer electrons (valence electrons); and sometimes, for the same reason, one atom, instead of sharing, gives an electron to another atom and a molecule is formed by *ionic* bonding.

Bonding of the oxygen molecule is shown in Fig. 3.4 and of the ozone molecule in Fig. 3.5.

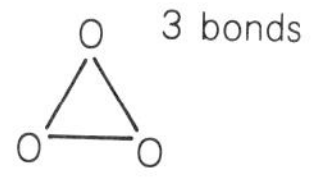

3 bonds

Fig. 3.5 The bonding of oxygen atoms to form an ozone molecule

When atoms pull together to form molecules, work is done which means that chemicals have energy — chemical potential energy.

Valency

The *valency* of an atom is a number that describes its combining power and is the number of arms the atom has (or, better, the number of *bonds* it usually forms). Oxygen has a valency of two (it is *divalent*), while a hydrogen atom has a valency of one (it is *monovalent*). Because oxygen is divalent we expect that, when it combines with hydrogen, each oxygen atom will bond with two hydrogens, as shown in Fig. 3.6, to produce the water molecule H_2O. Similarly, the carbon atom which has a valency of four (*tetravalent*), forms carbon dioxide by combining with two divalent oxygen atoms in Fig. 3.7.

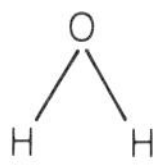

Fig. 3.6 The water molecule (H_2O)

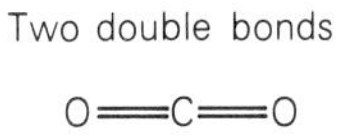

Two double bonds

$$O = C = O$$

Fig. 3.7 The carbon dioxide molecule (CO_2)

Elements, compounds and mixtures

An *element* is a substance made up entirely of atoms of one kind. For example, hydrogen, oxygen, carbon and iron are each elements.

A *compound* is a substance made up entirely of molecules of one kind, but each molecule is made up of more than one kind of atom. For example, water is a compound, all the molecules being H_2O.

The properties of a compound may be quite different from those of the elements made from the same atoms as illustrated in Fig. 3.8 (overleaf).

Common salt (*sodium chloride*) is a compound. Each molecule is formed from a sodium atom (Na) and a chlorine atom (Cl) and the compound is denoted by NaCl.

A *mixture* is composed of molecules or uncombined atoms of more than one type. The ingredients mixed together do not combine and so no new molecules are formed. The properties of the mixture are those of its ingredients (Fig. 3.9, overleaf).

Zinc (Zn) is the metal seen as the shiny rust-resisting coating on a new dustbin

Oxygen gas (O₂) is invisible and is essential for breathing

Zinc oxide (Zn) is a white powder used in make-up

When colouring and perfume are added to a cosmetic, such as a face powder, the ingredients retain their colour and smell and contribute these to the face powder which is a mixture.

Cereal is sweetened

Fig. 3.8   Zinc oxide is quite different from zinc and oxygen

Fig. 3.9   A mixture acquires the properties of its ingredients

Solids, liquids and gases

Solids consist of atoms or molecules close together and held in place by bonds between them. Examples are seen in Fig. 3.10. Not shown in the Figure is the fact that the particles are shaking about, i.e. vibrating.

(a)

4 bonds hold each atom

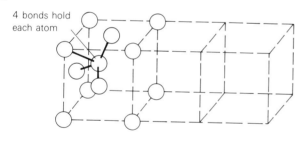

(b)

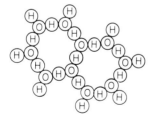

Fig. 3.10   The structure of a solid: (a) a tetravalent element; (b) ordinary ice

The bonding between the atoms or molecules of a solid frequently causes them to form a very tidy arrangement, and the solid is described as *crystalline*. This is true of the

two examples in Fig. 3.10. When a crystalline solid is broken the pieces produced have shapes that are related to the atom arrangement in the solid and these pieces are usually called *crystals*. Common salt and sugar are usually bought as small crystals.

In glasses, waxes and plastics the molecules hold together in a rather disorganised manner, so that these materials are *non-crystalline*.

Like a solid, a *liquid* consists of atoms or molecules close together. Here, however, the particles do not keep close enough together for bonding to prevent flow of the liquid. Fig. 3.11 suggests the rather disordered arrangement of molecules in water.

In a *gas*, the atoms or molecules are mostly well spaced out, attractive forces between them not being

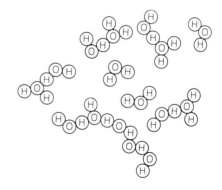

Fig. 3.11   Molecules of water (in liquid state) might be arranged like this

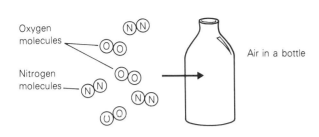

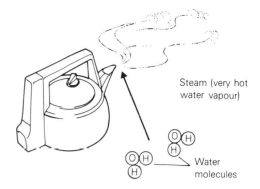

Fig. 3.12   Gas molecules are mostly well separated

noticeable. The molecules are continually 'rushing about' bumping into each other and the container walls, and they fill the whole of the space available to them (Fig. 3.12).

A *vapour* is, roughly speaking, an easily liquefied gas. It can be liquefied by squeezing it into a smaller space without cooling it. When a gas is obtained by evaporating a substance which is normally a liquid then the term *vapour* is usually appropriate. This is true, for example, of water and of ordinary alcohol (*ethanol*).

DENSITY

Density of a material

Density tells us how heavy a material is. Iron, for example, is a denser material than aluminium or poly-thene.

To be precise, the density of a material is the ratio of its mass to its volume or it is the mass per unit volume.

$$\text{Density} = \frac{\text{Mass}}{\text{Volume}}$$

A useful example to consider is water. 1000 cm³ weighs 1000 g so that the density works out to be 1000/1000 or 1 g per cm³. In g per m³ the value is 10^6 (or 1 million). Using the SI unit, its density is 1000 kg per m³.

Relative density

The *relative density* of a material, also called its *specific gravity*, is the ratio of the density of the material to the density of water. It is the number of times the material is denser than water.

The density of iron is 7800 kg per m³, and the density of water in kg per m³ is 1000, so that the relative density of iron is 7800/1000 or 7.8. Relative density is simply a number. It has no units.

Density and floating

Things of greater density than a liquid will sink in the liquid but things of smaller density will *float.*

Because of its lower density fat entering the stomach to be digested, floats on top of all the other liquid and consequently remains longest in the stomach.

When a solid object is floating in a liquid the volume of the solid which is above the liquid surface increases with the density of the liquid. This fact is employed in the

hydrometer method of measuring the density of a liquid. From the measured density of a solution the concentration (i.e. the strength) of the solution can be deduced, as suggested in Fig. 3.13 (overleaf).

Measuring densities can be useful, for example, to distinguish between different liquids which look the same, such as glycerine and liquid paraffin (approximately 1250 and 850 kg/m³ respectively).

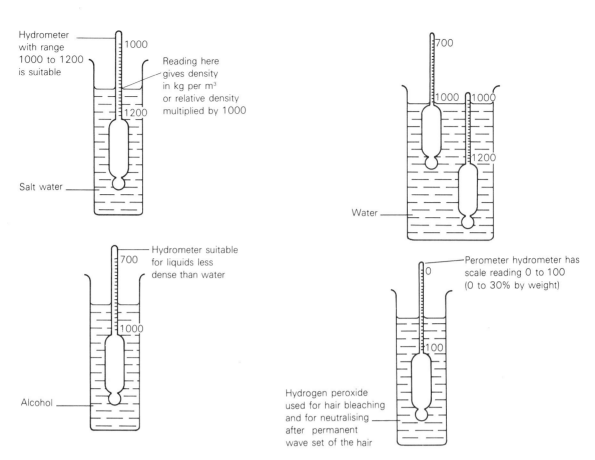

Fig. 3.13 The use of hydrometers

SUMMARY

- All matter is made up of protons, electrons and (except for hydrogen) neutrons.

- An atom consists of a nucleus containing protons and neutrons surrounded by electrons in orbit. The number of protons decides the type of atom, e.g. oxygen, hydrogen, iron, etc. The number of electrons in an atom normally equals the number of protons.

- Molecules are groups of atoms kept together by bonding forces between the atoms.

- The valency of an atom is its combining power.

- An element is a simple substance, all its atoms being of the same type.

- A compound consists of more than one type of atom. Its molecules are identical but each molecule consists of different types of atom. An example is water (H_2O).

- Solids consist of atoms or molecules close together and held in place by forces (bonding forces) between the particles.

- A gas has its atoms or molecules well spaced out.

- A liquid has its particles close together but free to move past each other.

- Density tells us how heavy a material is. Density is mass divided by volume.

- Density can be measured in g/m^3 or g/litre, or preferably, kg/m^3.

- Relative density (or specific gravity) is the density of material divided by the density of water.

- An object floats in a liquid if its density is less than that of the liquid.

- Density and relative density can be measured with a hydrometer.

1 An oxygen atom, oxygen molecule, electron, proton, neutron. Which is: (a) the smallest (has the smallest mass), (b) the biggest?

2. Metals, when in the solid or liquid state, are good reflectors of light (i.e. they are shiny if their surfaces are clean and smooth). They also allow electricity and heat to flow easily through them. Which of the following are *not* metals?

Brass, iron, aluminium, copper, nickel, carbon, glass, polythene

3. Which atom contains no neutrons?

4. In which part of an atom are the protons to be found?

5. What is the chemical symbol for: (a) the hydrogen atom, (b) the chlorine atom, (c) the sodium atom?

6. Which of the following are elements?
Hydrogen, oxygen, water, common salt, iron

7. Which of the following are mixtures?
Air, water, sodium chloride, face powder, lipstick

8. The nitrogen atom has five valence electrons. Explain why three hydrogen atoms and one nitrogen atom will combine by covalent bonding to produce ammonia whose chemical formula is NH_3.

9. Name one use in beauty therapy for each of the following:
(a) Carbon, (b) iron or steel, (c) zinc oxide

10. What is meant by the term *relative density*?

11. A certain piece of wax measures $10 \text{ cm} \times 20 \text{ cm} \times 5 \text{ cm}$ and it weighs 950 g. Calculate its density in g per cm^3. Also write down its relative density. Will this wax float or sink in water?

12. For what is a hydrometer used?

13. What volume of liquid paraffin is equivalent to 100 g weight? (relative density = 0.9).

14. A certain hydrometer floated in water with just its tip above the water surface. It floated with less of its stem submerged when placed in glycerine but it sank when placed in liquid paraffin. Explain these observations.

4

BASIC STATIC AND CURRENT ELECTRICITY

ELECTROSTATICS

Electric charge

A nylon blouse sometimes clings to the body and sparks as it is removed (Fig. 4.1). This is because it has acquired an *electric charge* by friction (rubbing).

Fig. 4.1   Sparks due to (static) electric charge

It is well known that small pieces of paper can be lifted up by a plastic comb or similar object that has been rubbed on a suitable cloth. The comb that has been rubbed is said to be electrified or electrically charged. It contains an electric charge.

Similarly an ebonite rod, if well rubbed on fur, will attract (and perhaps pick up) small objects such as pieces of paper or metal foil. A glass rod well rubbed on silk also shows that it is charged.

The effect is obtained only on the part of the rod that has been rubbed.

Two kinds of charge

If a charged rod, e.g. an ebonite rod, is held on a thread as shown in Fig. 4.2 (a) a similar charged rod will repel it. Repulsion also occurs if two glass rods are used.

However, if a glass rod is brought near to an ebonite rod (Fig. 4.2 (c)), the rods attract each other. (Note that these experiments are more easily done with other materials.)

Like charges repel. Unlike charges attract.

All materials can be electrically charged and are found to behave like ebonite *or* like glass (no charge ever being found that will attract or repel both ebonite and glass).

The charge possessed by ebonite rubbed on fur is called *negative* charge and the charge possessed by glass rubbed on silk is called *positive* charge. The signs — and + are used to denote negative and positive charge respectively.

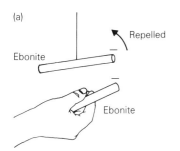 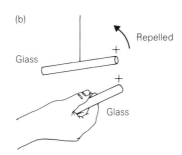 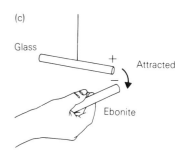

Fig. 4.2 Forces between charges

Explanation of charging

All materials, whether solid, liquid or gaseous, are made up from electrons, protons and neutrons, as explained in Chapter 3. Electrons are negatively charged, protons are positively charged, and neutrons are uncharged. The number of electrons found in each atom of any material normally equals the number of protons.*

A rubbed ebonite rod has more electrons than protons so that it repels another ebonite rod (Fig. 4.3).

A negative charge possessed by any object means a surplus of electrons. A positive charge means more protons, i.e. a deficiency of electrons.

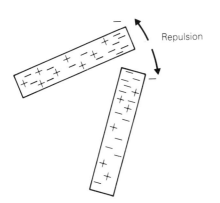 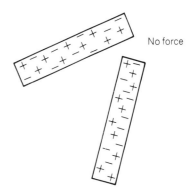

Fig. 4.3 The explanation of forces between charged rods

Electricity

What is *electricity?* The easiest answer to give is 'electric charge'. This is probably the best answer, too, because the various uses of the term electricity are all concerned with electric charges, their movements and their effects.

---

*All electrons are negatively charged and stay charged always, and every electron has exactly the same amount of charge almost as if someone had rubbed each one equally. Protons are exactly alike also, all being larger than electrons and all equally positive charged with just as much positive charge as the electron has negative charge.

Usually, materials contain equal numbers of protons and electrons. For example, an unrubbed ebonite rod has equal numbers of the two kinds of particle and does not repel a charged ebonite rod because, although its electrons repel the charged rod, its equal number of protons attract the negative rod equally.

Static electricity

When a comb or other object has been charged it can cause attraction or repulsion without the charge moving at all from the rod. In contrast, we shall see that heating by electricity or stimulating of muscle by electrical methods requires a movement of electric charge, i.e. it requires an electric current. For this reason we distinguish between *static electricity* (often simply called *static*) and *current electricity*.

Static electricity usually concerns very small charges (i.e. small surpluses or deficiencies of electrons) but the eagerness of the − and + charges to come together can cause charge movements through the air which are accompanied by small but visible and audible sparks.

Problems arising from static electric charges

Most people are aware of the attraction between a comb and the hair it has groomed, and equally troublesome is the phenomenon of hair refusing to lie down due to the repulsion between the similarly charged hairs (Fig. 4.4).

Fig. 4.4   Like charges repel and unlike charges attract

Hair shampoos will frequently leave the hair charged and this charge can be neutralised by the subsequent use of an appropriate hair conditioner.

Clothing can become electrically charged because, as the wearer moves about, the garment can acquire a charge by friction with adjacent material. On garments made from synthetic materials (man-made polymers like nylon) the charge builds up and is unable to escape. As a result clothing may cling to the wearer and a nylon overall may be attracted to nearby objects. Such effects can be annoying. Charged surfaces tend to attract dust. However, this effect is found to be useful in some air-cleaning equipment.

A hazard from electric charges can arise when you have been walking on a nylon carpet. If you then touch some metalwork (or perhaps even a client), you may experience a small electric shock. This is produced by a spark between your finger and the metalwork (Fig. 4.5). This spark, which may even be visible, is not dangerous because the amount of charge moving is very small, but it may startle you for a moment, perhaps causing an accident.

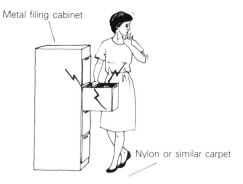

Metal filing cabinet

Nylon or similar carpet

Fig. 4.5   A slight shock from electrostatic charge can be most unpleasant

A further danger associated with sparks from electrostatic charges arises from the possibility of igniting highly flammable materials — for example, acetone (which is used in nail-varnish removers).

Metals, conductors and insulators

Metals are different from ebonite, glass and most other materials in that a large proportion of the electrons in a metal is always free to move around. Consequently any charge, i.e. extra electrons given to the metal, can spread through the metal. Metals are therefore called *conductors* of electricity.

The human body is also a conductor of electricity, i.e. it lets electric charge move through it, although not as easily as does a metal.

Metals are good conductors. The human body is a reasonably good conductor. Ebonite, glass, plastics and the air are *insulators* (they do not conduct electricity except under very special circumstances).

Charging by induction

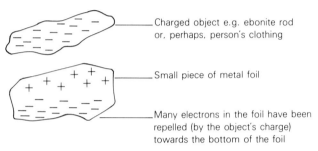

Charged object e.g. ebonite rod or, perhaps, person's clothing

Small piece of metal foil

Many electrons in the foil have been repelled (by the object's charge) towards the bottom of the foil

Fig. 4.6   A metal foil charged by induction

An object can be charged by friction and, if it is a conductor, it can be charged by connecting it to a battery or another charged conductor. But charging by *induction* means charging an object without having to touch it or connect it to anything. Fig. 4.6 shows an example of induction.*

CURRENT ELECTRICITY

Electric current

Electric *current* means a flow of electric charge.

When an electric current is flowing in a metal wire it means that electrons are moving along the wire, as illustrated in Fig. 4.7.

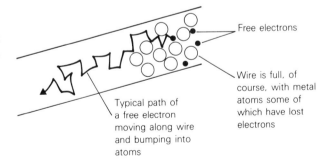

Free electrons

Wire is full, of course, with metal atoms some of which have lost electrons

Typical path of a free electron moving along wire and bumping into atoms

Fig. 4.7   The movement of a (free) electron along a wire

---

*In the figure, how does the size of the + charge on the top of the foil compare with the size of the negative charge at the bottom? What is the direction of the force between the object and the + charge? What is the direction of the force between the object and the other charge on the foil? Which of these forces is the greater? Your answers should show why there is an attraction between a rubbed ebonite rod and a piece of metal foil or small piece of paper. Similarly a nylon overall that has become charged by friction may be attracted to, and may cling to, metal furniture or to the person wearing the overall.

The unit for electric current measurement

This is the *ampere* (A). A current having a size of 1 ampere is a flow such that about 6 000 000 000 000 000 000 electrons are entering, passing along and leaving the wire or other conductor every second.

A milliampere (mA) is one thousandth of an ampere.

Potential difference (p.d.) between two places

The *potential* of a place may be described as its attractiveness or desirability for electrons.

A high potential place is a place that is attractive to electrons. This high potential could be due to the presence of a positive charge nearby, but usually it is simply due to there being a high concentration of positive charge (a shortage of electrons) at the place concerned. Similarly a low potential place is usually due to a lot of electrons being there.

With potentials of places we use the + and − signs to mean higher and lower potentials respectively. Thus, in Fig. 4.8, the piece of metal at one end of the cell is labelled as + and the metal at the other end − Electrons will move if they can from − to + potential places. Current flow requires a *potential difference* between two places and a conductor connected between these two places.

Current direction

It is usual to describe the direction of current flow as from + to −. As a result of this convention the current flow direction in a metal wire is opposite to the actual electron flow.

Electric cells

An electric *cell* or voltaic cell is a piece of apparatus that provides a potential difference. To do this it makes use of chemical action as explained below. These cells are better known by the name *batteries* although this term really should be used for a number of cells joined together. A *Leclanché dry cell* is shown in Fig. 4.8. This is the familiar cell that is used for torches, bicycle lamps, portable radios, electrical toys and electric door bells. It is not rechargeable.

A voltaic cell contains two different conductors as *electrodes* (for example, carbon and zinc) as shown in Fig. 4.8 (b), and a liquid (or paste) called an *electrolyte*.

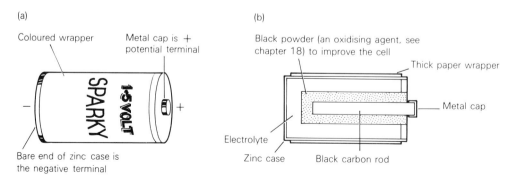

Fig. 4.8 The Leclanché dry cell: (a) external appearance; (b) internal construction

Chemical action in the cell makes one electrode become + charged and the other − charged, and it continues until a definite size of potential difference is achieved. This potential difference is called the *electromotive force* (e.m.f.) of the cell.

When the cell is not being used to produce a current (any circuit to which it is connected being *open*, i.e. incomplete), the chemical action in the cell will have no difficulty in providing the expected e.m.f. The p.d. between the terminals can be measured with a *voltmeter*, and the reading should agree with the expected e.m.f. as suggested in Fig. 4.9.

(a)

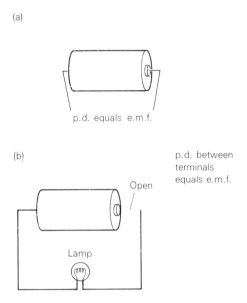

p.d. equals e.m.f.

(b)

Open

Lamp

p.d. between terminals equals e.m.f.

(c)

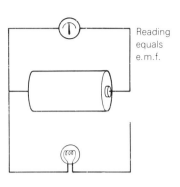

Reading equals e.m.f.

(d)

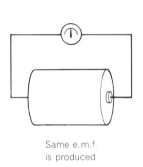

Same e.m.f. is produced

Fig. 4.9 A voltaic cell on open circuit:
(a) no connections to cell;
(b) circuit incomplete;
(c) a voltmeter is used to measure p.d. between terminals;
(d) a larger cell of same type

Current flow caused by a cell or battery

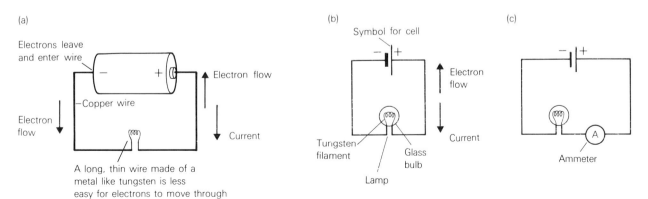

(a)

Electrons leave and enter wire

Electron flow

−   +

Electron flow

Copper wire

Current

A long, thin wire made of a metal like tungsten is less easy for electrons to move through

(b)

Symbol for cell

−   +

Electron flow

Current

Tungsten filament

Glass bulb

Lamp

(c)

−   +

Ammeter

Fig. 4.10 Electric currents: (a) producing an electric current; (b) simple circuit diagram; (c) an ammeter can be connected to measure current

When a complete circuit is connected to a cell, as in Fig. 4.10, a current flows. Electrons are repelled out of the negative terminal, pushing before them the free electrons in the copper wire. At the other end of the wire electrons are attracted in to the positive terminal. Thus electrons flow as shown, although we say current flows in the opposite direction. When the electrons arrive at the + terminal the cell removes them to keep this terminal + and keeps the other terminal − by adding electrons to it. The cell really acts as a pump, pumping electrons, and forcing them round the circuit.

Ammeters are used to measure electric current.

The unit for p.d.

Potential differences are measured in *volts* (abbreviation V). For example, the e.m.f. between (or 'across') the terminals of a Leclanché cell is 1.5 V. Two of these cells joined together in *series* would give 3 V, three would give 4.5 V, and so on. Fig. 4.11 shows cells in series and also in *parallel*. A lead–acid *accumulator* cell has an e.m.f. of approximately 2 V and is rechargeable. Six of these in series can be used to produce a 12 V car battery.

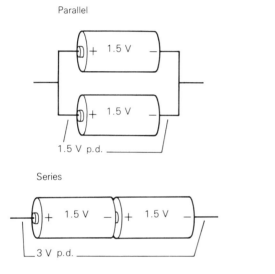

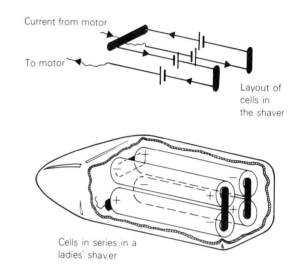

Fig. 4.11   Cells connected in series and in parallel

Power packs

Batteries have to be replaced as their essential chemicals are used up. Thus it is often more convenient and economical to use instead a piece of apparatus called a *power pack, battery eliminator* or *voltage supply unit,* that plugs into the mains electricity supply and provides the required voltage between two terminals.

The output may be low-voltage (low-tension, LT) or high-voltage (high-tension, HT), fixed voltage or variable. High voltages can be dangerous (see page 53).

Heating by an electric current

Wherever an electric current flows there is some heating.*

---

*We may picture the free electrons moving along a wire and bumping into the atoms of which the wire is made. This must make the atoms move, or move more, which means that the wire gets hot. Actually this movement of the atoms is a shaking movement, a vibration.

Heating particularly occurs in wires made of materials such as tungsten in a lamp bulb or Nichrome metal in an electric fire.

The heating is most noticeable where the wire becomes thin. Copper wires of reasonable thickness do not warm appreciably unless the current is very large. To obtain a lot of heat a considerable length of thin tungsten or Nichrome wire may be used.

Ohm's law

This law or rule states that the current flowing in a conductor, (for example in a metal wire) is proportional to the p.d. across it, provided that the temperature of the conductor does not change. It applies to most conductors.

The circuit shown in Fig. 4.12 illustrates Ohm's law because the current is found to double if the voltage used is doubled, and it trebles if the voltage is trebled.

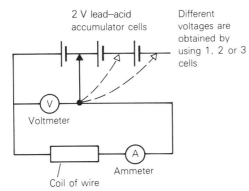
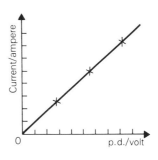

Fig. 4.12 Ohm's law

Resistance of a conductor

Another way of stating Ohm's law is to say that, while we experiment with the same conductor, voltage V divided by current I always gives the same answer. A different conductor, for example, another coil of wire, will also have a constant value for V/I, but the value for the second coil will be larger if it is more difficult for current to flow through it. The value of V/I tells us the *resistance* of the conductor to current flow.

The resistance of a conductor is its V/I value. Using the letter R for resistance gives an equation

$$R = V/I$$

where V is the p.d. across the conductor and I is the resulting current through it. If V is in volts and I in amperes, our answer for R is in *ohms* (abbreviation Ω).

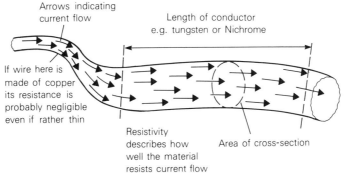

Fig. 4.13 Resistance and resistivity

For example, if a p.d. of 240 V produces a current of 5A, then the resistance of the conductor is

$$R = \frac{240}{5}$$

so that
$$R = 48 \ \Omega$$

The resistance of a conductor is decided by its length, its area of cross-section and by its *resistivity*. This is explained further in Fig. 4.13.

The resistance is greater for greater length, smaller cross-section and higher resistivity.

The resistivity of a metal like copper is small. It is larger for metals like tungsten and for mixtures of metals (*alloys*) like Nichrome and also for carbon (which is not a metal). Insulating materials like glass, plastics, etc., have extremely high resistivities. The resistivity of a metal can also be changed according to its temperature.

Resistors

A *resistor* is a device which has been made simply to provide resistance. Resistors are frequently coils of wire made from an alloy of suitable resistivity. Also, powdered carbon mixed with other powders and a binder can be packed in a small tube to make a resistor called a *radio resistor*.

A suitable resistor can be included in the electric circuit of a piece of beauty therapy apparatus in order to keep the current down to a safe value.

Dangerous electric currents

The electric currents discussed here have been *direct currents* (i.e. they have had constant flow directions). A direct current of perhaps 70 mA (70/1000 ampere), or possibly less, through the body could be lethal. Such currents may be produced by voltages perhaps as low as 70V DC.

Resistances in series

If a current flows as in Fig. 4.14 (a), through a resistance R_1 ohms and then through another resistance R_2 ohms, then the total resistance is $R_1 + R_2$ ohms.

For example if, as in Fig. 4.14 (b), there are two 48 Ω resistors in series, then the total resistance is $R_1 + R_2$ which is $48 + 48$ or $96 \ \Omega$. Therefore the current expected is $I = \frac{12}{96} = \frac{1}{8}$ ampere.

(a)

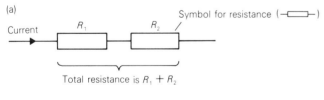

Current — R_1 — R_2 — Symbol for resistance (—⊏⊐—)

Total resistance is $R_1 + R_2$

(b)

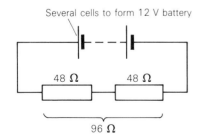

Several cells to form 12 V battery

48 Ω 48 Ω

96 Ω

(c)

MERRY XMAS TO OUR CLIENTS

e.g. 20 lamps, 48 Ω each, total resistance 960 Ω

Fig. 4.14   Resistances in series

Resistors in parallel

Two resistors connected in parallel are shown in Fig. 4.15. The resistance of this combination is equal to R_1 multiplied by R_2 and then divided by $R_1 + R_2$.

If, for example, a resistance of $2\,\Omega$ is required for a circuit but only $3\,\Omega$ and $6\,\Omega$ resistors are available, a parallel combination is the answer.

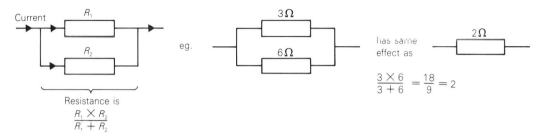

Fig. 4.15 Resistors in parallel

Resistivity of body tissues

Impure water, and therefore moist body tissues, will conduct noticeably.

Dry skin has a rather high resistivity. Wet skin has a lower resistivity. Fat beneath the skin is quite a good insulator while muscle tissue conducts quite well, its resistivity being rather low (see Fig. 4.16).

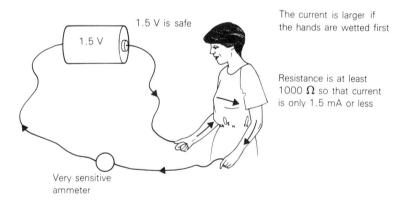

Fig. 4.16 The conduction of electricity through the body

Internal resistance

It is common to find that, inside a voltaic cell, there is some resistance, not deliberately built into it but an unavoidable part of its construction. In fact, it is the resistance of the electrolyte between the electrodes, through which charge has to be transferred, as explained earlier.

For example, a 1.5 V dry cell measuring about 2 inches in length and of about 1 inch in diameter will often have

an internal resistance — of about $1\,\Omega$. When it is connected to a $5\,\Omega$ resistance the current is given by

$$I = V/R = 1.5/6 = \tfrac{1}{4} \text{ ampere}$$

The p.d. actually across the $5\,\Omega$ is also less than expected as a result of the presence of the internal resistance. The p.d. across the $5\,\Omega$ is given by

$$V = I \times R = \tfrac{1}{4} \times 5 = 1\tfrac{1}{4} \text{ volts}$$

As shown in Fig. 4.17, internal resistance can cause the current obtained to be less than expected and also the useful p.d. to be less than expected.

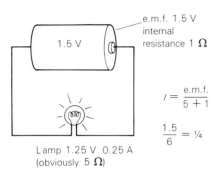

e.m.f. 1.5 V
internal resistance 1 Ω

1.5 V

$$I = \frac{e.m.f.}{5 + 1}$$

$$\frac{1.5}{6} = \frac{1}{4}$$

Lamp 1.25 V 0.25 A
(obviously 5 Ω)

Current meter

Even with conductors touching an excessive current does not flow, due to internal resistance in the apparatus

Apparatus to produce voltage for electro-chemical hair removal ('galvanic' method)

Fig. 4.17   The effects of internal resistance

Electric power

When an electric current is moving through a conductor it does work. Usually this is because free electrons, as they try to move along, push upon, and move, the atoms that get in their way. The atoms, in fact, vibrate and their kinetic energy makes the conductor warmer.

The amount of heating produced each second will depend upon the 'eagerness' with which the electrons are trying to get through the conductor and the number moving through per second. It is not, therefore, too surprising that the work done per second (*power*) or heat produced per second is given by the product of p.d. times current:

Power $P = V \times I$

As explained in Chapter 1, the unit for power is the watt (W). Thus we can say

Watts = Volts $\times$ Amperes

As an example, a tungsten filament lamp (Fig. 4.18(a)) rated at 60 W and designed to work on a 240 V supply will, when used, have flowing through it a current given by

$$I = \frac{P}{V} = \frac{60}{240} = \frac{6}{24} = \frac{1}{4} \textbf{ ampere}$$

We have assumed here that a direct current is being used but, as explained later, an alternating current (AC) may be used instead.

It may be useful to note that, since $P = VI$ and $V = IR$, then

$P = I^2R$, and, as an alternative, $P = V \times \dfrac{V}{R}$ i.e. $\dfrac{V^2}{R}$

$$\therefore \quad P = I^2R \text{ or } \frac{V^2}{R}$$

(a)

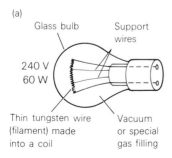

Glass bulb

Support wires

240 V
60 W

Thin tungsten wire (filament) made into a coil

Vacuum or special gas filling

(b)

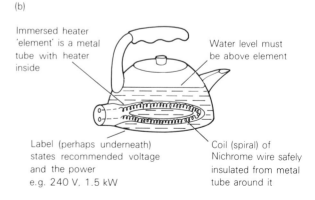

Immersed heater 'element' is a metal tube with heater inside

Water level must be above element

Label (perhaps underneath) states recommended voltage and the power e.g. 240 V, 1.5 kW

Coil (spiral) of Nichrome wire safely insulated from metal tube around it

Fig. 4.18   Electrical power: (a) tungsten filament lamp;   (b) electrically heated kettle

ALTERNATING CURRENTS (AC)

The meaning of AC

A *direct* current has a constant direction of flow, but an *alternating* current is one that flows first one way round the circuit and then the opposite way and keeps reversing over and over again. This is shown in Fig. 4.19.

An alternating current is produced by a voltage of alternating polarity.

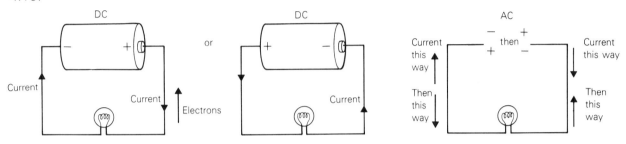

Fig. 4.19   Direct and alternating currents

Different types of alternating current

An alternating current could have a constant size, for example, 2 A, for one-thousandth of a second (1 ms) and then suddenly this flow is stopped and a flow of 2 A is produced in the opposite direction. A second possibility is the ordinary AC (also called *sinusoidal*) which gradually increases to a maximum (or *peak*) value, then decreases to zero and then, in the opposite direction, increases and decreases before reversing direction again. These differences are most easily displayed by sketching graphs as in Fig. 4.20.

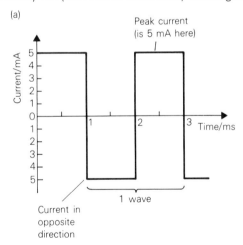

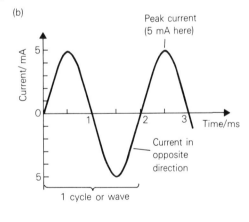

Fig. 4.20   Two examples of alternating currents: (a) a square wave;   (b) a sine wave (sinusoidal)

Cycle of alternating current

A *cycle* is one complete alternation, including flow one way and flow the opposite way, i.e. 1 wave in Fig. 4.20.

Frequency of alternation

Frequency (*f*) tells us the number of complete cycles that occur every second. We can write

$$f = \frac{\text{Number of cycles}}{\text{Time taken for this number of cycles}}$$

The unit for frequency is the *hertz* (abbreviation Hz). 1 Hz means 1 cycle per second. The frequency of the ordinary mains supply is 50 Hz in the United Kingdom.

The *period* is the time taken for 1 complete cycle to occur.

To describe the size of an alternating current

The simplest way of describing the size of an alternating current is to state the peak current. For ordinary, sinusoidal AC a more useful figure is the peak current multiplied by 0.707 because this value of the current (called the *RMS value*) is needed for most calculations. Similarly the peak value of an alternating voltage (often called, inappropriately, AC voltage) is not as useful as the RMS voltage (= 0.707 × peak voltage).

If RMS values are used for currents and voltages, the methods for AC calculations are the same as for DC.

Unless otherwise indicated, meters for measuring ordinary alternating currents and voltages give RMS answers.

Mains AC supply in the United Kingdom is 240 V RMS and is sinusoidal.

ELECTRIC CURRENTS FOR BEAUTY THERAPY

High-frequency, galvanic and faradic currents

Electric currents of carefully controlled size are made to flow through parts of a client's body for a variety of reasons, as described in later chapters (Fig. 4.21). For some treatments, such as hair removal by the diathermy method (Chapter 13) and for high-frequency skin treatment (also Chapter 13), an alternating current of very high frequency is employed. For iontophoresis and disencrustation techniques (see Chapter 5), a direct current is essential. A direct current used for medical or beauty therapy is often called a *galvanic* current and the use of such a current is called *galvanism*. Electric currents are also used to stimulate muscles, i.e. make them operate, and to stimulate nerves, either so that the nerves cause muscles to operate or so that they cause increased blood supply to the part of the body treated. For those two purposes a current which changes quite quickly is needed and, in beauty work, the currents are

made to change suddenly as explained below. This kind of current is called *faradic* and its use in therapy is called *faradism*. Fig. 4.22 shows the differences between the common types of current used for therapeutic purposes.

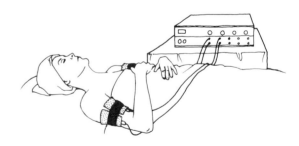

Fig. 4.21 Electric currents being used for beauty therapy — direct (galvanic) current being used with the object of reducing 'fat' (or cellulite, see p. 44)

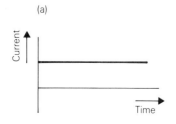

(a)

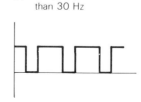

(b) Frequency less than 30 Hz

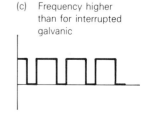

(c) Frequency higher than for interrupted galvanic

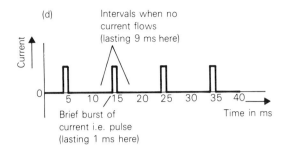

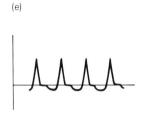

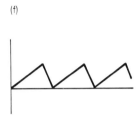

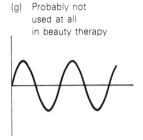

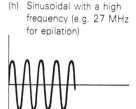

Fig. 4.22   Types of current: (a) galvanic current;   (b) interrupted galvanic;   (c) rectangular wave;   (d) typical example of modern faradic current;   (e) original faradic current;   (f) triangular wave;   (g) sinusoidal; (h) high-frequency current (approaching 1 MHz or higher)

Surging of faradic currents

A faradic current of the kind shown in Fig. 4.22 (d) can cause the muscle or muscles affected to pull tight, i.e. contract, and stay contracted. The 1 ms pulse duration is ideal for healthy muscles, a shorter time necessitating larger currents and a longer time offering no advantage. The period is 10 ms (frequency 100 Hz), which is not too fast for muscles to respond. However, two further considerations should be made. First is the fact that a muscle will tend to react less and less to a succession of

identical pulses, and second, that the muscle should rest. For these reasons the faradic current can be modified as illustrated in Fig. 4.23.

A typical faradic voltage supply is shown in Fig. 4.24 (overleaf).

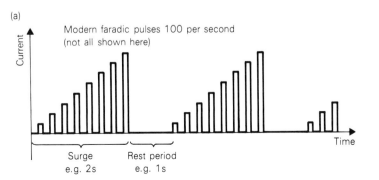

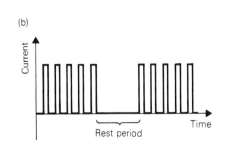

Fig. 4.23   Modified faradic current: (a) surging faradic current;   (b) interrupted faradic current

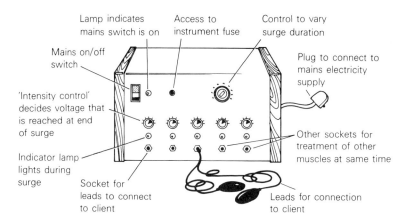

Lamp indicates mains switch is on

Access to instrument fuse

Control to vary surge duration

Mains on/off switch

Plug to connect to mains electricity supply

'Intensity control' decides voltage that is reached at end of surge

Indicator lamp lights during surge

Socket for leads to connect to client

Other sockets for treatment of other muscles at same time

Leads for connection to client

Fig. 4.24 An example of a faradic voltage supply unit (with surging)

Instrument fuses

A *fuse* is a thin piece of wire made of a metal that melts quite easily. If as the result of a fault an abnormally large current flows that could damage a supply unit or the instrument, this fuse should quickly melt ('fuse' or 'blow') and so cause an open circuit and an end to the current. A fuse in the mains plug (see Chapter 6) may provide adequate protection but instrument fuses are common.

When the fault has been remedied a new fuse is fitted. Occasionally a fuse will blow in the absence of a fault.

Some simple electrical tests

To check that the wire inside a fuse or inside a lamp is not broken, i.e. there is *continuity*, we can use a circuit as shown in Fig. 4.25 (a). If no current is obtainable then the wire is broken and a new item must be purchased.

To check that a voltage is being supplied at an output socket on a galvanic unit, a DC voltmeter can be used (Fig. 4.25 (b)). This should preferably have a choice of ranges, for example, 0 – 100 V, 0 – 10 V, 0 – 1 V, so that the highest voltage range can be tried first, then the next range and so on, until a useful reading is obtained. In this way there is no danger of applying too large a voltage for any meter scale, and if the wrong polarity of connection is made to the meter (the pointer will move below zero on the scale), this is discovered before harm is done and the connections to the meter can be interchanged.

To examine a voltage from a galvanic, faradic or high-frequency unit a *cathode ray oscilloscope* can be used (Fig. 4.25 (c)) and this instrument actually shows the waveform of the voltage, which may of course agree with any of the graphs shown in Fig. 4.22 (pp 36–7).

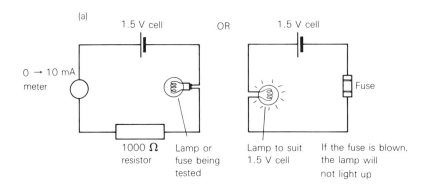

(a)

1.5 V cell   OR   1.5 V cell

0 → 10 mA meter

1000 Ω resistor

Lamp or fuse being tested

Lamp to suit 1.5 V cell

Fuse

If the fuse is blown, the lamp will not light up

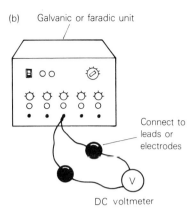

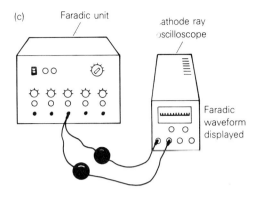

Fig. 4.25 Some electrical tests

Battery-operated equipment

The advantage of battery-operated equipment is that it can be used anywhere, but the disadvantage is the inconvenience and expense of replacing batteries.

The e.m.f. of a suspect cell or battery can be checked as suggested in Fig. 4.9 (c). If it is noticeably lower than its normal value, the cell is probably of no further use.

Even if the e.m.f. is apparently normal, the internal resistance may be high if the cell is weakening. If the terminal p.d. can be measured, by use of a suitable voltmeter, of course, while the cell is in use in the equipment, then a lower than normal reading shows the cell or battery is faulty.

Rechargeable batteries can be bought. These can be recharged many times, the procedure being to send a current through the battery in the opposite direction, i.e. into its + terminal and out of its − terminal. The current size and time required for recharging should be ascertained from the supplier.

Safety of electrical treatments

When electric currents are used for therapy purposes great care is required not only to get good results, but to ensure maximum client comfort and safety. These matters are discussed in Chapters 6 and 13.

Short circuits

If the terminals of a voltage supply are connected by a very low resistance the circuit is described as a *short circuit*. The terminals are said to be *shorted*.

This is illustrated in Fig. 4.26 where current flows very easily along the path shown by the arrows. The resistance of this path, formed only of copper connecting-wire, is very low. The current through the lamp is almost zero because it is so easy for the current to go from + to − without going through the lamp. The lamp does not light. The current is 'taking a short-cut to its destination. We have a *short circuit*.

When a short circuit occurs, the current may be huge.

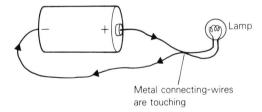

Fig. 4.26 A short circuit

Overheating of a wire, a cell or materials touching the wiring can result in melting, fire or other damage.

Unless it is known that the internal resistance of a voltage supply is high enough to prevent too large a current, we must avoid producing a short circuit.

Insulated conductors

To guard against short circuits it is usual to enclose (or wrap) connecting-wires in a flexible, insulating cover.

Cotton and enamel are sometimes used, but plastics such as PVC (see Chapter 18) are often used for this purpose.

SUMMARY

- Electrostatics is the study of charges at rest (i.e. not moving as a current all the time).

- The word *electricity* is used to mean electric charge.

- An object that can noticeably repel or attract another object yet is not a magnet is said to be electrically charged.

- Like charges repel; unlike charges attract.

- Electrons are negatively charged; protons are positively charged.

- Normally an object has equal numbers of protons and electrons, and therefore it is uncharged (or neutral).

- A negatively charged object has a surplus of electrons; a positively charged object has a deficiency of electrons.

- Electrostatic charges produced by friction are small but they can be a nuisance:
  o Hair is attracted to the comb.
  o Hairs may repel each other.
  o Clothes may cling.
  o Sparks can startle you or even cause fires.

- Metals are examples of good electrical conductors.

- Examples of electrical insulators are glass, rubber, most plastics, gases like air.

- Carbon and alloys (mixtures of metals) are quite good conductors.

- Impure water and moist body tissues conduct noticeably.

- An electric current is a flow of charge. In a metal wire it is a flow of electrons.

- The unit for electric current is the ampere (A).

- Potential is the attractiveness of a place for electrons.

- A difference of potential between two places is needed if a current is to flow.

- Potential difference (p.d.) is measured in volts.

- The + and − of a voltaic cell are the higher potential and lower potential parts of the cell respectively.

- The direction of a current is said to be opposite to the electron flow direction and so the current flows from + to − of a cell.

- A battery is a number of cells joined together.

- A single cell produces a potential difference of only about 1 or 2 V depending on its type.

- For a current to flow there must be a p.d. between two places and also a conductor joining these places. Usually this means that there must be a complete circuit.

- An electric current produces heating in a conductor given by:
  Heat (in joules) = Current × p.d. across
  per second   conductor

- Ohm's law states that the current doubles if the p.d. (voltage) doubles, provided that the same conductor is used and the temperature does not change.

- The resistance of a conductor tells us its opposition to current flow. Its exact meaning is p.d. divided by current:

$$R = V/I$$

  The unit for resistance is the ohm (Ω).

- A resistor is a device made for the purpose of providing resistance.

- Resistances in series are added to obtain the total resistance.

- Resistances in parallel have a combined resistance given by

$$\frac{R_1 \times R_2}{R_1 + R_2}$$

- Resistivity tells us about a material. A material that current easily flows through has a low resistivity. Dry skin has a high resistivity compared with moist tissue like muscle. Body fat does not conduct very well.

- Some voltage supplies, dry cells, for example, have some internal resistance.

- A direct current of 70 mA or possibly less could be lethal depending upon the path it takes through the body.

- Electric power (i.e. the electrical energy converted into heat or some other energy form, per second) is given by

$$P = V \times I$$

and is measured in watts (W).

- An alternating current (AC) is one that keeps reversing its direction repeatedly and regularly. An alternating voltage is required to cause it.

- The frequency of an alternating current is the number of times it repeats per second. 1 cycle (or repetition) per second is called 1 hertz (Hz). The time taken for one cycle to occur is called the period.

- Galvanic current is DC used for therapy or medical purposes.

- Faradic current is a pulsed current of frequency greater than 30 Hz.

- Mains supply (ordinary AC) has a frequency of 50 Hz and is sinusoidal.

- High-frequency currents have a frequency approaching 1 million hertz or higher.

- An instrument fuse protects the instrument in case too large a current flows through it.

- A short circuit is a circuit which has too low a resistance and, in some cases, it can cause excessive current to flow.

EXERCISE 4

In questions 1 – 9 state which of the alternatives suggested is correct.

1. A current/p.d. is a flow of electric charge.

2. A current in a metal wire consists of a movement of electrons/protons along the wire.

3. All materials contain protons/electrons.

4. Electrons will try to spread out away from each other/will attract each other.

5. (a) Electrons are always +/− charged.
(b) Protons are always +/− charged.

6. The force between two like charges is a repulsion/attraction.
The force between two unlike charges is a repulsion/attraction.

7. An uncharged object has equal numbers of protons and electrons/a preponderance of neutrons/no electrons or protons.

8. When an uncharged object (e.g. small piece of metal foil) is close to a + charged rod, electrons in it move so that it becomes +/− on the side nearest to the rod and +/− on the other side. The force on the side nearest to the rod is directed towards/away from the rod and is bigger/smaller than the force on the opposite side. For this reason the object is weakly attracted/repelled by the rod.

9. A well-rubbed rod of ebonite repels another ebonite rod/attracts a glass rod well rubbed on silk/weakly attracts an unrubbed rod or other uncharged object.

10. Give two undesirable effects of static electricity in the salon. (CGLI)

11. State the SI units for: (a) current, (b) potential difference, (c) power.

12. What is meant by the e.m.f. of a voltaic cell?

13. What e.m.f. is obtained from three 1.5 V cells connected together (a) in series, (b) in parallel?

14. What e.m.f. is expected if two voltaic cells are connected in series but with the two + terminals touching?

15. What is the resistance of a circuit which has an e.m.f. of 100 V and produces a current of 5 amps?

16. What voltage is needed across 4 lamps in series if each requires 6 V across it?

17. Two $10\,\Omega$ lamps each requiring 6 V across it are connected in parallel. What p.d. must be applied across the combination and what current will flow through each lamp?

18. What is the resistance obtained by connecting a $20\,\Omega$ resistor in parallel with a $5\,\Omega$ resistor?

19. Which of the following is the best conductor of electricity: dry skin, muscle, body fat?

20. Name two common electrical insulating materials. (CGLI)

21. Calculate the current that flows in a 40 W lamp when it is connected as recommended to a 240 V supply.

22. A small electric heater contains two $60\,\Omega$ heating coils (elements). When a switch is set at (a) 'high', the elements are connected in parallel, when at (b) 'medium', one element only is used and, when at (c) 'low', the two elements are connected in series. If the supply voltage is 240 V what is the current in each case, and how much heat per second is produced in each case?

23. Distinguish between the following types of electric current: (a) alternating, (b) direct, (c) high-frequency. (CGLI, part question)

24. Give the meaning of: (a) frequency, (b) period.

25. Name one use for a galvanic current in beauty therapy.

26. (a) Distinguish between galvanic and faradic currents.
(b) Why are faradic pulses usually of 1 ms duration?

27. What is the function of an instrument fuse?

28. What is meant by surging of a faradic current?

29. Name one use for high-frequency electric current in beauty therapy.

5

GALVANIC CURRENTS AND ELECTROLYSIS

GALVANIC CURRENTS

Beauty therapists usually employ this term to describe a direct electric current that is being passed through the body for any kind of therapy. Strictly speaking, *galvanic* refers to an electric current produced by chemical action.

Electrolysis

This is the process which accompanies the flow of direct current through body tissue whereby chemical changes are caused in the vicinity of the electrodes. *Electrolysis* also occurs when a direct current flows through various solutions of chemicals in water.

Use of galvanic currents in beauty therapy

In the field of beauty therapy it is for hair removal (*epilation*) that galvanic current and electrolysis are best known. In fact, epilation is often misleadingly called electrolysis. Today, however, it seems that epilation by electrolysis is rarely used because it is a slow method compared with the more modern *diathermy* method compared with the more modern *diathermy* method which destroys the base of the hair root by heating (*thermolysis*).

However, iontophoresis and disencrustation therapies are common, and they use galvanic currents.

A demonstration of genuine electrolysis

Fig. 5.1 (overleaf) shows how a solution of potassium iodide may be made to conduct. The solution is initially colourless but an impressive brown colour soon appears near the positive electrode indicating that a chemical change has occurred. The brown colour is due to the production of iodine. Chemical change is also found near the negative electrode. If a piece of red litmus paper is introduced close to this electrode the litmus turns blue. This is a consequence of the production of potassium hydroxide.

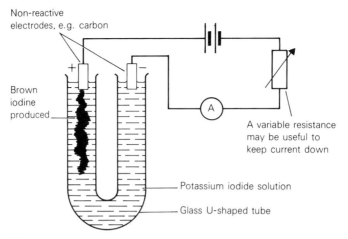

Non-reactive electrodes, e.g. carbon

Brown iodine produced

A variable resistance may be useful to keep current down

Potassium iodide solution

Glass U-shaped tube

Fig. 5.1   A demonstration of electrical conduction in a solution

Electrolyte

When electrolysis takes place, the electrical conduction is not due to a simple flow of electrons through the conducting material but involves the movement of ions as explained below. The word *electrolyte* is used for the solution or other material in which this conduction occurs. A solution of sodium chloride in water (i.e. a solution of common salt in water) is the electrolyte of importance for our discussion. The liquid in body tissues is largely sodium chloride solution.

Electrodes

The conductors used for making contact with the electrolyte are called *electrodes*. Some electrodes used in beauty therapy are shown in Fig. 5.2. Metals, being good electrical conductors, are the usual choice for making electrodes for galvanic treatments.

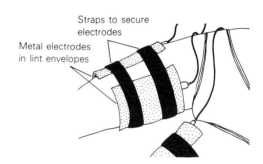

Straps to secure electrodes

Metal electrodes in lint envelopes

Fig. 5.2   Electrodes applied to thigh for galvanic treatment of cellulite

Ionisation

Atoms or molecules that are charged as a result of adding to them or taking from them one or more electrons are called *ions*. The process of creating ions is known as *ionisation*.

The molecule of sodium chloride ionises as shown in Fig. 5.3 to produce a positive sodium ion and a negative chlorine ion. This ionisation takes place as the salt is being dissolved in the water so that ions are

44

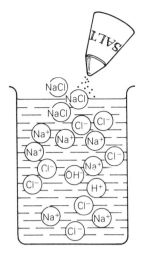

Fig. 5.3 The formation of ions

present before any current flow is attempted One way of representing ionisation in symbols is this:

$$NaCl \longrightarrow Na^+ + Cl^-$$

Sodium atom + Chlorine atom Sodium   Chlorine

                                                         ion   ion

Sodium chloride molecule

Also some water molecules will become ionised, an H_2O molecule splitting into a positive hydrogen ion (H^+) and a negative hydroxyl (or hydroxide) ion (OH^-).*

Conduction by ions

Conduction in an electrolyte occurs when a potential difference is produced between the electrodes (e.g. a battery is connected). The positive ions in the electrolyte then move towards the negative electrode (called the *cathode*) because positive particles are attracted to a negative place. The negative ions move to the positive electrode, called the *anode*. This is shown in Fig. 5.4.

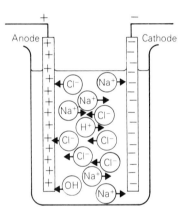

Fig. 5.4 Conduction by ions

Results of ion movements in sodium chloride solution in body tissues

At the cathode sodium hydroxide (caustic soda) is produced in the electrolyte and hydrogen gas is released as bubbles.†

At the anode, chlorine ions (Cl^-) arrive, give up their charge and produce hydrochloric acid and hypochlorous acid in the electrolyte. Some oxygen gas may sub-

---

*Actually hydrogen ions so formed always join up with water molecules to produce hydroxonium ions ($H^+ + H_2O \rightarrow H_3O^+$), but this need not concern us here.

†A simplified (and therefore not perfectly accurate) explanation of this result is as follows. The sodium ions reach the negative electrode and obtain electrons from it so that each sodium ion is neutralised and becomes a sodium atom. However, the hydrogen ions in the adjacent electrolyte then take the electron from the sodium atom so that ultimately it is the hydrogen ions that are neutralised and form hydrogen gas.

$$Na^+ + OH^- + H^+ \longrightarrow Na^+ + OH^- + H^+$$

            Water   Sodium hydroxide

and H^+ + **Electron from cathode** $\longrightarrow$ H atom $\xrightarrow[\text{in pairs}]{\text{combine}}$ **Molecules of hydrogen gas**

sequently be released. Some production of chlorine gas is also a possibility.

We note that the electrolyte close to the cathode has become alkaline due to a growing concentration of hydroxyl ions while, at the anode, the electrolyte becomes acid. The current flow is largely due to the movement of sodium ions (Na^+) towards the cathode and chlorine ions (Cl^-) towards the anode.*

GALVANIC TREATMENTS

Epilation using electrolysis

In this technique the negative electrode is in the form of a needle, often made of stainless steel, which is inserted in the follicle of the hair to be removed. The positive electrode, wrapped in moistened material such as lint, may be gripped by the client's hand (see Fig. 5.5).

Fig. 5.5   Galvanic epilation

The epilation needle enters the hair follicle as shown in Fig. 5.6. When current is made to flow it is the caustic soda produced around the cathode needle that destroys the base of the hair root and follicle. A current of about 0.2 mA is likely to be suitable. A time of about a minute may be required. The current size is probably decided entirely by the voltage used and by the resistance within the voltage supply unit so that the area of the needle in contact with the hair follicle will largely influence only the spreading of the current over this area. Thus a thinner needle may be less comfortable for a given current due to the caustic soda being concentrated over a smaller area. It should be noted that the comparative dryness, and hence poor conductivity of the outer epidermis, is fortunate in that little conduction and chemical action will occur here. If, instead of this technique, a hot needle were moved into a hair follicle to achieve epilation, it would first burn the top layer of the epidermis. Galvanic epilation is slow but effective.

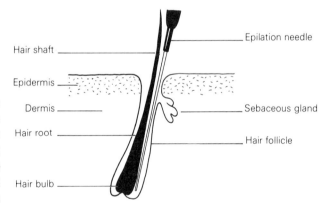

Fig. 5.6   The epilation needle

Testing for electrode polarity

Two tests for electrode polarity are shown in Fig. 5.7.

If the two electrodes are placed in some sodium chloride solution and the voltage applied, then it is at the cathode that bubbles of gas will first be noticed.

If a piece of red litmus paper or similar testing paper is moistened with sodium chloride solution, and the electrodes are applied to this, then it is at the cathode that a blue colour will be seen.

---

*The electrons given by the chlorine ions to the anode then move through the wire from the anode as expected of a current flowing in a metal wire. In the wire connected to the cathode electrons move towards the cathode where they will be taken up by the sodium ions being neutralised there.

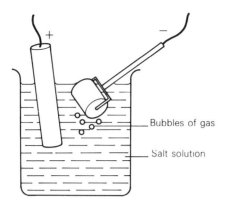
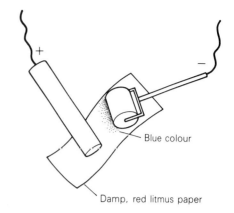

Bubbles of gas

Salt solution

Blue colour

Damp, red litmus paper

Fig. 5.7   Testing electrode polarity

Iontophoresis

Iontophoresis is the name given to the technique whereby ions of a chosen chemical are made to enter the unbroken skin by placing a solution containing these ions beneath an electrode applied to the skin. If the ions concerned are positive (called *cations* because they move towards the cathode), then the electrode under which the solution is placed (called the *active electrode*) should be positive so that the ions are repelled into the skin. The other electrode (called the *passive* or *neutral* electrode) may be held in the client's hand. This process which causes cations to be moved is often called *cataphoresis*. If it is required to move negative ions (*anions*) into the skin (*anaphoresis*), then the solution of these ions is placed beneath the electrode applied to the skin but this must now be made the negative electrode.

Iontophoresis is used for introducing nourishing substances deeply into the skin, particularly on the face. Fig. 5.8 shows a roller electrode being used on the face. This electrode is probably made of plated steel. For treating smaller areas of the face the active electrode can take the form of a pair of tweezers holding some cotton wool soaked in the special iontophoresis solution.

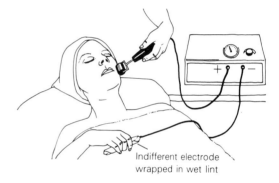

Indifferent electrode wrapped in wet lint

Fig. 5.8   Iontophoresis treatment of the face using a roller electrode

A further use is made of iontophoresis when it is desired to reduce *cellulite*, which is excessive 'soft fat' beneath the skin (see Chapter 20) and which gives the skin a bumpy appearance. The aim here is to introduce as ions a substance that will eliminate the cellulite. When such treatment is applied to the body the electrodes will normally consist of sheets of easily shaped metal, often tin.

Disencrustation

Disencrustation is a means of removing surface oiliness, due to sebum and other unwanted matter from the skin, by employing electrolysis. This is achieved by using galvanic current, as in iontophoresis. A special solution may be used for this purpose; it is placed beneath the active electrode. Alternatively, salt solution may be used beneath the electrode and in this case the polarity required is negative.

Electrode padding

The active electrodes used in iontophoresis (and dis-encrustation) are, except for roller and tweezer electrodes, separated from the skin by a layer of lint or similar material which is soaked in the chosen solution. The passive electrode is wrapped in similar material soaked usually in water. The functions of the padding are to soften, i.e. moisten the skin so as to reduce its electrical resistance, to improve contact between the electrode and the skin without any part of the metal electrode touching the skin and to absorb excess caustic products of the electrolysis.

Other electrode effects

At the positive electrode the skin becomes drier and firmer, nerves are soothed, and blood vessels are constricted somewhat. At the negative electrode tissues are softened, irritation may be felt and redness is seen because of the *vasodilation* (widening of the blood vessels). The sensations usually felt are a tingling beneath the cathode followed by a feeling of warmth. Too large a current causes discomfort, and so too does a sudden change of current. For example, if contact is accidentally lost between an electrode and the body or if the voltage supply is switched off, there will be an unpleasant electric shock. This in turn may lead to sudden muscle contraction and further accidents.

Some of these electrode effects can be demonstrated on a piece of meat (Fig. 5.9).

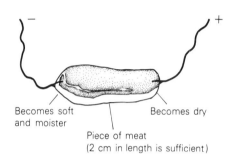

Fig. 5.9 A demonstration of some electrode effects

Interrupted galvanic currents

Interrupted galvanic currents already described in Chapter 4 can be used medically for stimulating muscles with damaged nerve supplies, but these currents are not used in beauty therapy. It is the sudden changes in current due to its being interrupted that cause the muscle contractions.

Reversal of electrode polarity

The chemical effects of a current are reversed if the polarity of the electrode is changed. For this reason it is common practice to reverse the polarity of the electrodes at the end of some galvanic treatments in order to reduce the alkalinity in the skin that has been treated.

Alternating current that flows first in one direction through the body and then equally strongly in the opposite direction will produce no chemical effect.

Safety

Supply voltages of up to 120 V and currents of many milliamperes may be used, for example, in iontophoresis, but, provided that the current is increased and decreased very steadily by careful adjustment of the supply unit's voltage control (probably marked as 'intensity' control), then there should be no danger to the client. The resistance of the body plus resistance built into the supply unit will limit the current to a safe value. Of course, the paramount rule is that the therapist slowly increases the current and does not exceed the current acceptable to the client.

Number and area of electrodes

With the usual galvanic supply for iontophoresis the effect of connecting a second active electrode on the one supply (the connection being arranged in parallel — see Chapter 4) is to increase the total current taken from the supply with little reduction of the current to the first electrode. Only one passive electrode need be used, as shown in Fig. 5.10.

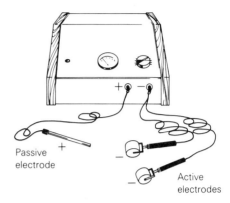

Passive electrode

Active electrodes

Fig. 5.10 Connections for two active electrodes, e.g. roller electrodes

SUMMARY

- A direct current used for cosmetic or medical purposes is called galvanic current.

- Electrolysis means producing chemical changes by use of an electric current.

- A direct current is needed for electrolysis.

- Galvanic epilation is slow but effective.

- Atoms that have an extra electron or are short of an electron are called ions — negative or positive ions respectively.

- Ionisation is the production of ions and occurs when some chemicals dissolve.

- Sodium chloride dissolves to give sodium ions and chlorine ions.

- An ionised solution will conduct electricity.

- Conduction through body tissues is largely due to the presence of sodium chloride dissolved in the tissue fluids.

- To send a current through a solution or through body tissues, electrodes are used, the + one being called the anode and the − one being called the cathode.

- When current flows through sodium chloride solution (for example, in body tissue), sodium hydroxide (an alkali) is produced at the cathode and hydrogen gas. At the anode, acid is produced and some oxygen gas.

- Epilation by electrolysis uses a needle that is placed in the hair follicle and acts as the cathode. The sodium hydroxide (caustic soda) produced around the needle attacks the hair root and surrounding follicle.

- When testing the polarity of galvanic electrodes by immersion in salt solution (sodium chloride), bubbles of gas first appear at the cathode.

- The cathode produces blue coloration of damp red litmus paper.

- Iontophoresis uses galvanic current to send chemicals through the skin.

- Disencrustation, by use of galvanic current, removes surface grime.

- At the anode the skin is dried. At the cathode it is softened and reddened and may be irritated.

- Beware of sudden changes in galvanic current, for example, breaking contact. It makes muscles contract, sometimes violently. Always change current (intensity) gradually.

1. Distinguish between:
   (a) anode and cathode
   (b) electrode and electrolyte
   (c) iontophoresis and ionisation
   (d) ion and iron.

2. What particles are responsible for current flow in: (a) a metal, (b) a solution of common salt in water? In which directions do the particles mentioned move?

3. What is the chemical responsible for destroying the base of the hair root when epilation is obtained by use of galvanic current?

4. What word is frequently used by beauty therapists in place of the word *epilation* even when the epilation is not achieved by a chemical but by a heating effect of an electric current?

5. Describe one way of checking the polarity of a pair of electrodes connected to a galvanic (voltage supply) unit.

6. Why does chemical action around an epilation needle destroy tissue yet the same amount of chemical action beneath an electrophoresis electrode produces a much milder affect?

7. Give one reason for wrapping padding material such as lint around an electrode.

8. What polarity is needed for the active electrode when using salt water for disencrustation?

9. Why does chemical action around an epilation needle, using the galvanic method, not destroy the outer end of the hair follicle?

10. What precautions must be taken to avoid discomfort or harm to the client when iontophoresis therapy is being carried out?

11. State two effects of an electric current and describe, in detail, the practical application of one of these in the treatment of superfluous hair. (CGLI)

12. What is the effect of passing a DC current through a solution of salt water? (CGLI)

6

MAINS ELECTRICITY

MAINS SUPPLY

The National Grid

If many homes, factories and other buildings are connected to a common supply, they are said to be connected to a *supply main*. They have a mains supply.

In the United Kingdom not only are most buildings connected by a complex wiring system to power stations, but there are additional connections between power stations, with the result that the whole country is covered by a single network of conductors called the National Grid.

Socket outlets

Many people are justifiably interested only in the mains supply where it reaches them in the very room that they occupy. They see the *socket outlets* on the wall, or lamp sockets fitted to the ceiling. These sockets are drawn in Fig. 6.1. Electrical equipment (or appliances) needing mains electricity supply is plugged into the sockets.

Live (L), neutral (N) and earth (E) conductors are explained later in this chapter.

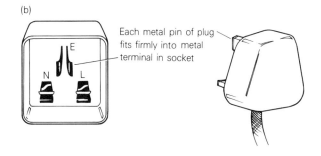

(b)

Each metal pin of plug fits firmly into metal terminal in socket

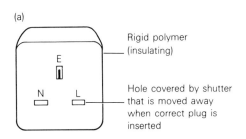

(a)

Rigid polymer (insulating)

Hole covered by shutter that is moved away when correct plug is inserted

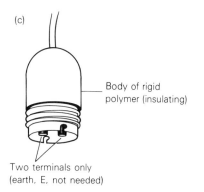

(c)

Body of rigid polymer (insulating)

Two terminals only (earth, E, not needed)

Fig. 6.1   Electricity sockets: (a) appearance of a typical socket outlet;   (b) socket outlet with face removed (after safely disconnecting the mains supply) and plug to fit;   (c) lamp socket

Mains voltage supply

In the United Kingdom the voltage supplied, that is the potential difference provided between the L and N terminals of a socket, is '240 volts, AC 50 Hz'.

As explained in Chapter 4, AC means that we have an alternating voltage that will produce an alternating current (AC). Thus, at one moment, L will be positive compared with N. Then the polarity will be reversed so that L is the negative of the two terminals. Then the polarity will reverse, and repeat the cycle 50 times per second. 50 hertz (50 Hz) is the frequency.

The mains voltage alternates as shown in Fig. 6.2 and reaches as much as 340 V but the voltage quoted is approximately seven-tenths of this and is the RMS* value discussed in Chapter 4. This value is 240 V.

The advantage of using the figure of 240 V instead of the peak value of 340 V is that it gives correct answers for mains electricity calculations using the same formulae as for DC calculations. This has already been pointed out in Chapter 4.

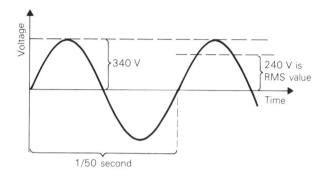

Fig. 6.2   Mains supply voltage

Payment for electricity supplied

It is the energy supplied that decides the cost. The energy, in joules, is equal to the power in watts (i.e. joules per second) multiplied by the time (in seconds) for which it is used:

Energy in joules = Watts × Seconds

The cost of electrical power is, however, usually given for the use of 1000 watts (1 kW) for 1 hour. This amount of energy is called a *kilowatt-hour* or *electricity unit*. The tariff will be stated as, perhaps, 5 pence per unit.

As an example, 1.5 kW electric heater running for 4 hours would use 1.5 × 4 kilowatt-hours and cost, at 5p per unit, 6 × 5 or 30 pence. (The number of joules used is, in fact, 1500 (watts) × 14 400 (seconds) i.e. 21 600 000). A typical electricity meter is shown in Fig. 6.3.

When we speak of the 'cost of electricity' we use the word 'electricity' to mean 'electrical energy'. When we say 'switch on the electricity' we mean 'switch on the voltage (potential difference)'.

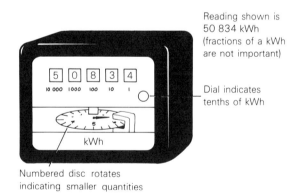

Fig. 6.3   A typical electricity meter

---

*RMS stands for root mean square. It is the square root of the average value of the voltage squared.

HAZARDS AND PRECAUTIONS

The hazards of mains electricity

The word *hazard* means a possible danger and, indeed, there will always be a hazard when we are working with voltages exceeding a few volts, especially with mains AC voltages. See Fig. 6.4!

However, if we take the proper care, then the danger will remain only a possibility and not a reality.

Hazard without precautions = Danger!

If a few milliamperes, or more, of 50 Hz electric current passes through the body, a very painful electric shock may be felt and, depending upon the route the current takes through the body, muscles contract and stay contracted — a condition called *tetanus*. This can stop the victim from letting go the faulty electrical equipment responsible for the shock. Breathing may become difficult and may stop, and death from exhaustion may follow unless the current is soon stopped. Larger current will cause severe burns and, if it passes through the chest, will cause *fibrillation* of the heart. Fibrillation means that the pumping action of the heart becomes disorganised and death, due to inadequate blood supply to the brain, can be expected within minutes whether or not the current is stopped.

Injury shock also accompanies electric shocks. This means that the amount of blood in circulation is reduced, leading to pallor and coldness of the skin and in severe cases this shock can cause collapse and death (see Chapter 23).

Quite apart from electric shock there is the hazard of fire being caused by the heating effect of an excessive electric current, caused usually by a short circuit.

Fig. 6.4 Electric shock. No sensible person would do this; but accidental shocks occur only too easily

Hazards of direct currents

The main difference between the effects of 50 Hz AC and direct or galvanic current is that the 50 Hz AC causes continuous contraction (i.e. tetanic contraction) of muscles. Direct current only causes muscle contraction when the current starts or stops. A person with a direct current passing through, say, his or her hand and arm can open the hand to let go if the pain of doing so can be tolerated.

A second difference is that a larger current can be tolerated without harm when the current is direct.

It must be emphasised that AC and DC can both be lethal.

The earth

It is useful to know that:

(a) the earth on which we live is not an insulator. It is not difficult for large currents to flow through the earth.

(b) The earth is so huge that its potential is effectively unchanged by any charge which we might send into it.

Live and neutral conductors

For technical reasons, one of the two wires which brings electricity to the user in the United Kingdom is connected to the earth before it reaches the consumer's premises. This *neutral* wire and the neutral terminal to which it connects in each socket are always equally attractive to electrons. The potential of the neutral terminal cannot change much.

The *live* (L) terminal is alternately more attractive for electrons than the neutral (i.e. L is +) and then less attractive than the neutral (i.e. L is −). It is the potential of the live terminal that alternates.

The earth terminal (E) is provided for safety reasons as we shall soon see.

A mains supply is shown in Fig. 6.5.

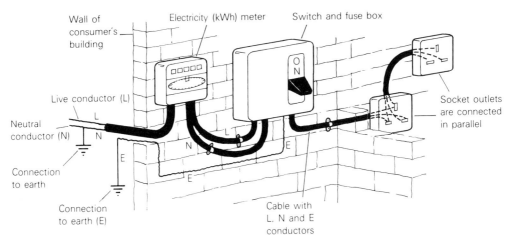

Fig. 6.5   Mains supply

Earth shock

Electric shocks in which alternating current flows through the body between the live and neutral conductors, as shown in Fig. 6.4, are not the commonest type of shock. Far more often shock is caused by current that flows between a live conductor and the earth, passing through the body in doing so. This is shown in Fig. 6.6.

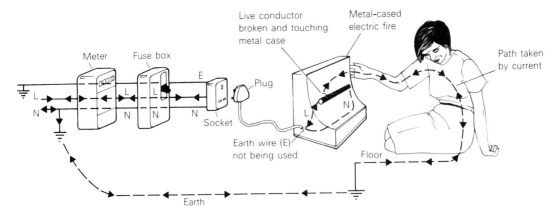

Fig. 6.6   Earth shock — what could happen if the earthing wire (E) is not connected to the appliance (an electric fire here)

Protection against earth shock by use of an earth wire (E) and a suitable fuse

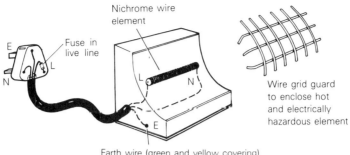

Fig. 6.7   An electric fire correctly fitted

If an *earth wire* (E) is connected (as shown in Fig. 6.7) and a fuse is fitted in the live lead (usually a cartridge fuse, Fig. 6.8, fitted in the plug), then many electric shocks can be prevented. If the live wire ever touches the metal case, a huge current will flow through the case and the earth wire and the fuse will blow in the live lead. This disconnects the appliance from the live supply so that it is safe to touch.

It is essential that the fuse used should blow as soon as the current becomes greater than normal.

The normal current is given by

$$I = \frac{\textbf{Wattage of appliance}}{\textbf{Supply voltage}}$$

For a 2 kW fire:

$$I = \frac{2000}{240} = 8\tfrac{1}{3} \ \textbf{ampere}$$

Given a choice of (say) 1, 2, 3, 5, 10 or 13 A fuses, the one that blows at 10 A would be suitable for the 2 kW fire.

The fuse in a plug also protects the appliance to which it is connected by limiting the size of current that can flow should a fault occur.

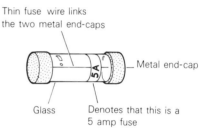

Fig. 6.8   A cartridge fuse suitable for a mains plug

All-insulated appliances

These are surrounded everywhere by an insulating case so that there is no need to earth the case.

Because the exposed parts of electric lamp bulbs and tubes are made of glass they are connected to live and neutral terminals only.

Double-insulated appliances

A square inside a square
Earthing is not necessary

Fig. 6.9   The symbol for double insulation

A metal-cased piece of equipment may have extra insulation fitted inside to ensure that live wires cannot touch the metal case. Such apparatus is marked by a special symbol (Fig. 6.9) and earthing is not required.

British Standards specifications

The British Standards Institution provides nationally recognised standards for quality and safety of equipment and for technical procedures. Their recommendations are mostly published as written standards each of which is identified by a number preceded by the letters BS. For example, BS 3456 describes the safety requirements of electric heaters and similar appliances and BS 415 applies to electronic equipment.

The BSI is also concerned with approval of equipment, checking that it complies with the appropriate British Standard. Equipment that is of approved design is marked as indicated in Fig. 6.10.

Another organisation that checks to see that appliances conform to the relevant British Standard is the BEAB (British Electrotechnical Approvals Board).

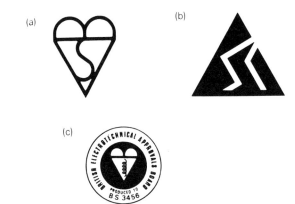

Fig. 6.10   Design and safety marks:
(a) BSI kitemark;   (b) BSI safety mark;
(c) BEAB safety mark

Fitting an appliance correctly to a plug

First, one must obtain suitable flexible cable. Assuming that the appliance requires earthing, the cable must have L, N and E conductors, and the current rating of the cable must be adequate for the normal operating current of the appliance to be carried without any problem.

The colours of the insulation on the wires should be brown for L, blue for N, and green and yellow stripes for E, as shown in Fig. 6.11. (Some years ago the colours used were red for L, black for N and green for E. Cable on very old appliances may have these colours but the cable should have been replaced by now!)

The correct rating for the fuse is calculated or found from the appliance manufacturer's instructions and the fuse is fitted into the plug.

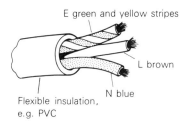

Fig. 6.11   The colour code for mains flexible cable

With the aid of a screwdriver and suitable wire-strippers/cutters, the plug is fitted to the cable as illustrated in Fig. 6.12. Note that the plug itself must be one in which the terminals can handle a current up to the fuse rating.

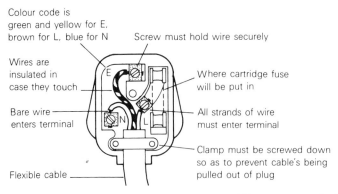

Fig. 6.12   The wiring of a mains plug, e.g. a 13 ampere rated plug

Mains fuse and switch box

This box is found where the mains electricity supply cable enters the building and is usually close to the electricity meter (see Fig. 6.3, p. 53). A typical box is illustrated in Fig. 6.13.

The purpose of each of the fuses in the fuse box is to protect the wiring between the fuse box and the socket outlets. The rating of each of these fuses must not exceed the current-handling capacity of the cable that it protects.

A circuit that provides outlets for lighting only may need to carry no more than 5 A. A circuit providing many sockets for delivering power to electric heaters, washing machines, a television set or beauty therapy voltage supply units usually requires a 30 A fuse.

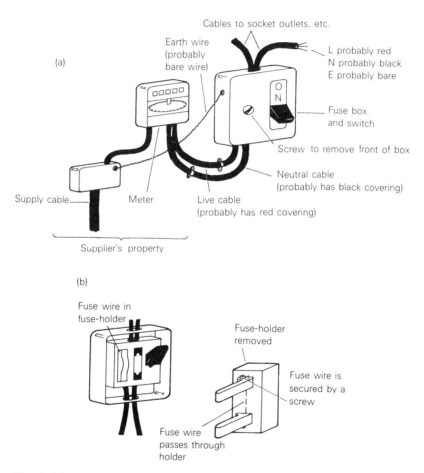

Fig. 6.13   A mains fuse and switch box: (a) general view;   (b) box opened

Isolating socket outlets

If an electrician is to fit or repair a socket outlet, he can disconnect the wiring to the socket from the electricity supply by switching off at the fuse box (to be safe he should remove the appropriate fuse too). Then he should check that the circuit concerned is off (perhaps by connecting a lamp to a socket in the circuit to see that it does not light up).

If you ever suspect a fault in the wiring behind or inside a socket outlet, you should switch off at the fuse box and send for an electrician.

Circuit-breakers

As an alternative to the use of a mains fuse, a *circuit-breaker* may be fitted which automatically throws the mains supply switch to the off position if excessive current flows. No melting of a fuse wire is necessary. An advantage here is that, as soon as the fault is repaired, the circuit can be brought into action again by simply resetting the switch. This switch may be of the press-button type.

Water and electricity

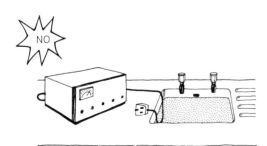

Fig. 6.14 Keep water and electricity apart!

Keep them apart!

Extra hazards arise, and these can be very serious, when water is present in the vicinity of high-voltage electrical equipment (Fig. 6.14). Wet electrical apparatus, the touching of electric sockets or switches with wet hands, having electrical equipment within reach of a person in a bath: each of these is a *disgrace*, certainly a terrible hazard, perhaps a crime under the Health and Safety at Work Act, and possibly the cause of death or appalling injury. Correct wiring and earthing and correct choice of fuse may no longer provide safety when water is involved.

The danger of electric shock from wet apparatus is illustrated and explained in Fig. 6.15. Current can flow between live and neutral by passing through the live wire, the water, the body and the earth.

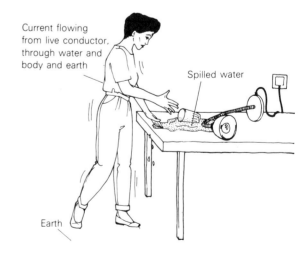

Fig. 6.15 Electric shock arising from a water spill

Earth-leak trips (ELTs)

Extra protection against dangerous electric shocks, particularly in damp situations, can be obtained with an *earth-leak trip*. If some current (a few mA or more) flows via the earth instead of by the proper route through the neutral wire then the earth-leak trip immediately disconnects the supply. An ELT is shown in use in Fig. 6.16. Often an ELT is included in the mains supply fuse box or fitted to it.

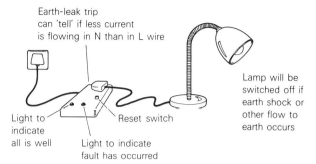

Earth-leak trip can 'tell' if less current is flowing in N than in L wire

Lamp will be switched off if earth shock or other flow to earth occurs

Light to indicate all is well

Reset switch

Light to indicate fault has occurred

Fig. 6.16 An earth-leak trip in use

Some electrical do's and don'ts

Don't use:

- an incorrect fuse or any kind of improvised fuse
- trailing cables which someone could trip over
- multiple adaptors which might cause the total current taken to exceed the rating of the socket outlet and wiring supplying it (Fig. 6.17)
- light sockets for purposes other than lighting
- electrical equipment near water
- worn cables
- plugs not fully pushed into sockets
- cables placed where they could become trapped in a door or drawer or under furniture or become melted by heaters

Do:

- carry out regular checks of location and condition of equipment and cables
- check plug wiring

- hold plugs by the insulating case only
- dry hands before touching plugs or equipment

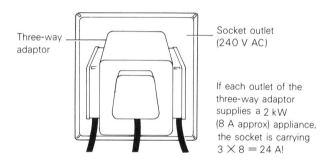

Three-way adaptor

Socket outlet (240 V AC)

If each outlet of the three-way adaptor supplies a 2 kW (8 A approx) appliance, the socket is carrying 3 × 8 = 24 A!

Fig. 6.17 A three-way adaptor can lead to overloading of a socket outlet

SUMMARY

- Nearly all buildings in the UK obtain electricity by connection to the National Grid.

- Socket outlets and plugs have live, neutral and earth terminals.

- Mains supply in the UK is 240 V (RMS) and its frequency is 50 Hz.

- The cost of electricity supplied is decided by the number of kilowatts multiplied by the number of hours of use, i.e. the number of kilowatt-hours (called electricity units).

- Mains (50 Hz) AC is particularly hazardous. Quite small currents through the body cause muscle contraction and the muscles are kept contracted (tetanus) as long as the current lasts, and the person may be unable to let go. Exhaustion can then soon lead to death. Currents through the chest are most likely to cause death. These currents can also cause pain and shock. Breathing may become difficult or stop. Large currents cause severe burns.

- The commonest electric shock is 'earth shock' caused by connection of the person between a live conductor and the earth.

- By earthing the metal case of an appliance and using the correct fuse in the live lead, earth shock can be avoided in nearly all cases.

- Use of the correct fuse also protects the appliance.

- The correct fuse must be rated at just a little more than the normal appliance current.

$$\text{Normal current} = \frac{\text{Power of appliance in watts}}{\text{Voltage}}$$

- The fuse for an appliance is usually a cartridge fuse fitted inside the plug.

- All-insulated apparatus has no exposed metal parts and earthing is not essential.

- Double-insulated apparatus does not need earthing.

- For joining a plug to an appliance the wires (flexible conductors) must have insulating covering of the following colours:

    Live — brown, neutral — blue, earth — green and yellow.

    (Previously the colours were: Live — red, neutral — black, earth — green.)

- When fitting a plug the wires must be: correct colours to L, N and E terminals of the plug, firmly screwed down, without loose strands, bare wires not touching each other. The cable must be secured where it enters the plug and the correct fuse must be fitted.

- The wiring behind the socket outlets is protected by a fuse in the mains fuse box or by a circuit-breaker.

- The mains fuse box is fitted with a switch to turn off the supply to sockets to which it is normally connected.

- Water near mains electricity is a serious hazard. An earth-leak trip gives greater protection against electric shock and is particularly advantageous where dampness can occur.

EXERCISE 6

1. State two differences between the voltage supplied by a Leclanché dry cell and by a mains socket outlet.

2. Name the three terminals of a socket outlet.

3. Name the two terminals of a lamp socket.

4. What is meant when the UK mains supply is described as 50 Hz?

5. Calculate the AC (or RMS) current that flows when 240 V RMS is connected to a 100 ohm apparatus.

6. Define the term *hazard*.

7. What are the two main hazards associated with the use of electricity in the salon? (CGLI)

8. Give two differences between the hazards of DC and AC current passing through the body.

9. State in one or two words the reason for having an earth connection to the metal frame or case of an electrical apparatus.

10. (a) Are socket outlets normally connected in series, parallel, both or neither?

(b) What effect on the voltage at a socket does one expect to find when another socket on the same circuit is brought into use?

(c) What current flows through a fuse-box when six outlets are in use, each supplying 2A?

11. What is the purpose of a fuse-box fuse?

12. What are the two functions of a fuse in a plug connected to an electric appliance such as a galvanic supply unit? (See also Chapter 4.)

13. State the colour of: (a) the live wire, (b) the earth wire in a three-core flexible cable. (CGLI)

14. Which of the following statements is/are correct?

(a) Tap-water is an insulator

(b) Water increases the risk of electric shock

(c) An electric shock current of more than a few milliampere flowing through the body to the earth would make an earth-leak trip switch off the supply.

15. State two safety precautions concerning the removal of a plug from a socket.

16. State two safety precautions concerning the placing of flexible cables.

17. Why should use of multiple adaptors generally be avoided?

18. How does mains current differ from modern type faradic current?

19. What is meant by a *double-insulated* appliance?

20. What faults can you see in the wiring of the plug shown in Fig. 6.18?

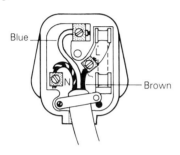

Fig. 6.18 Faults in the wiring of a plug

21. Describe the structure of a cartridge fuse. (CGLI, part question)

22. What type of electric current is supplied by: (a) a battery, (b) the mains supply? (CGLI)

23. (a) For what purpose is the term kilowatt-hour used?
(b) State the alternative name for this term. (CGLI)

24. What would be the cost of operating a 4 kilowatt steam bath for 10 hours at 3p per unit? (CGLI)

7

HEAT AND TEMPERATURE

TEMPERATURE

Hot and cold

How hot? How cold? What is the *temperature*?

These questions mean the same because temperature means 'hotness'. Well, perhaps we had better be a little more precise. Temperature is hotness measured in a reliable way, i.e. by use of a thermometer. High temperature means hot; low temperature means cold.

We are all familiar with the impression of some water feeling cold to the hands if one's hands were previously in hot water, yet the same lukewarm water feels warm if the hands were previously in very cold water. We ourselves are unreliable judges of hotness. So we use thermometers to measure temperatures (Fig. 7.1).

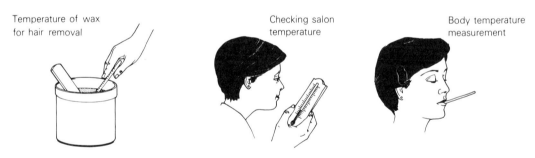

Temperature of wax for hair removal

Checking salon temperature

Body temperature measurement

Fig. 7.1 Measurements of temperature

Temperature scales

A method of measuring temperature is called a *scale*.

The *Celsius* scale (also called *centigrade*) is the method in which we describe as *zero* the temperature at which water normally freezes, and *boiling point* (or BP) the temperature at which water normally boils. The BP is called 100 degrees. (The unit for temperature measurements has always been called *degree*.) The normal freezing temperature (or *freezing point*, FP) for water is called the *ice point*, and the normal temperature at which water boils

(under normal atmospheric pressure) or, better, the temperature of the steam immediately above the boiling water, is called the *steam point*.

Denoting Celsius by C and degree by °, we have:

Normal FP of water $= 0°C$ ⎫
Normal BP of water $= 100°C$ ⎬ a gap of 100°C
Normal body temperature $= 37°C$ ⎭

The Fahrenheit temperature scale

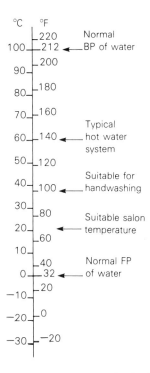

Fig. 7.2 Some temperatures in the beauty salon

With this scale the ice point is called 32 degrees and the steam point 212 degrees. Between these two temperatures there are 180 degrees, whereas on the Celsius scale there are only 100 degrees between the ice and steam points. Thus we see that the *Fahrenheit* degree is the smaller one.

Denoting Fahrenheit by F and degree by °, we have:

Lowest temperature
obtained by
Daniel Fahrenheit = 0°F
Normal FP of water = 32°F $\}$ a gap of 180°F
Normal BP of water = 212°F

Normal body temperature = 98.4°F

Temperatures below zero on either the C or F scale are described by the number of degrees counted below zero and this number is written with a minus sign. For example, a typical domestic refrigerator will have a temperature inside of about −5°C, meaning that the temperature is 5 Celsius degrees below the ice point. The normal temperature of a domestic freezer is −18°C (0°F). Some temperatures in the salon are shown in Fig. 7.2.

Conversion of a temperature from °C to °F and vice versa

For converting a temperature in °C to °F the procedure is as follows. The number of degrees Celsius is multiplied by $\frac{180}{100}$ (or by $\frac{9}{5}$) to obtain the number of degrees Fahrenheit counted from the ice point. 32 is then added so that the answer is the number of degrees F counted from the Fahrenheit zero. The procedure may be written as a formula:

$$\textbf{Number of } °\textbf{F} = \left(°\textbf{C} \times \frac{9}{5}\right) + 32$$

For conversion from °F to °C we must first subtract 32 from the F temperature to obtain the number of degrees Fahrenheit counted from the ice point. This number is then multiplied by $\frac{5}{9}$ to obtain the final smaller answer for the C temperature. The formula is:

$$°\textbf{C} = (°\textbf{F} - 32) \times \frac{5}{9}$$

To illustrate the use of these formulae consider the following two examples.

(a) The temperature of a sauna might be from 180 to 220°F. Suppose a certain sauna is providing a temperature of 203°F, what is this temperature in °C?

$$°\textbf{C} = (203 - 32) \times \frac{5}{9} = 171 \times \frac{5}{9} = \frac{855}{9} = 95$$

The answer is 95°C.

(b) Consider the approximate melting point of beeswax that is being used in making some cosmetics. Suppose this is given as 65°C. What is this temperature in °F?

$$°\textbf{F} = \left(65 \times \frac{9}{5}\right) + 32 = (13 \times 9) + 32 = 117 + 32 = 149$$

The answer is 149°F.

HEATING AND EXPANSION

Heating

Heating means 'raising the temperature' of something.

Cooling means 'reducing the temperature'.

When something is heated the microscopic movements of its atoms and molecules (vibrations, etc.) are increased.

Thermal expansion

An increase of volume usually occurs when something is heated. Cooling is usually accompanied by a volume decrease. Heating expands; cooling contracts.

The volume changes are very small for solid objects, usually bigger for liquids, and can be much bigger for gases.*

A gas always occupies the whole of its container, but it will try to expand when it is heated because its molecules move faster, hit the walls of the container harder and try to enlarge the container. In other words, the pressure on the inside of the container walls is increased by heating.

Most metals when heated expand quite a lot compared with other solids and so they are described as having a comparatively high *thermal expansivity*.†

Solids have a fixed shape so that thermal expansion means an increase in length, breadth and depth.

Thermal expansion (and contraction) can be very useful for measuring temperatures and for controlling temperatures (see thermometers and thermostats below).

Uneven expansion or contraction of solids, such as glassware and crockery items that are suddenly heated or cooled, frequently causes cracking. Pouring very hot water into the bottom of a glass jam-jar will cause the bottom to expand and break away from the cooler sides of the jar. Using borosilicate glass instead of ordinary soda glass makes cracking less likely due to the lower thermal expansivity of borosilicate glass. It is, however, bad practice to place hot glass or ceramic containers on cold surfaces.

Two examples of thermal expansion are shown in Fig. 7.3.

Hot water pouring over a metal bottle-cap can make it expand and loosen

Uneven heating and thermal expansion of glass has led to the hazards of escaping scalding water and sharp glass edges

Fig. 7.3 Examples of thermal expansion

---

*If atoms or molecules keep reasonably still they can hold close together. If a piece of metal, for example, is heated its atoms vibrate more and take up more space. When the metal is cooled the vibrations are reduced so that the attractive forces between the atoms are better able to pull the particles together.

† A long rod or bar of metal might expand by 1 part in 100 000 for each Celsius degree of temperature rise (its thermal expansivity 1/100 000 or 10^{-5} per °C), so that 1 metre of rod will, for say 100 degrees of temperature rise, expand by 1 mm.

Bimetal strips

These are strips made up of two lengths of different metals riveted together as shown in Fig. 7.4. The two metals are chosen to have quite different thermal expansivities so that the heating causes one metal to expand more than the other and bend in doing so.

Bimetal strips are used for making certain thermometers and thermostats, some of which are described below.

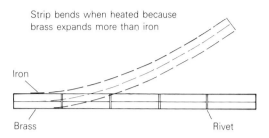

Fig. 7.4 A bimetal strip

THERMOMETERS AND THERMOSTATS

The mercury-in-glass thermometer

Fig. 7.5 shows a typical thermometer of this type.

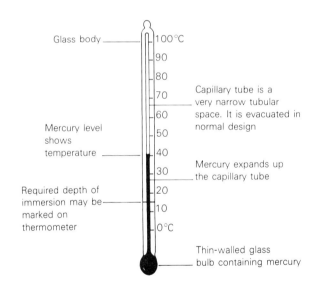

Fig. 7.5 A mercury-in-glass thermometer

This thermometer makes use of the thermal expansion of mercury which, when heated, expands a little but rises a lot in the narrow (capillary) tube. The mercury level is quite easy to see.

Heat entering the thermometer spreads quickly through the mercury (which is a good conductor of heat — see Chapter 10) and the temperature reading is obtainable in usually much less than a minute.

To measure the temperature accurately the thermometer should be immersed to the required depth. Some thermometers are calibrated (i.e. the degree markings are put on) so that the thermometer needs to be totally immersed to obtain correct readings.

The range of this type of thermometer may be 0° to 100°C, as in Fig. 7.5, or anywhere between −40°C and 360°C (the freezing and boiling points of mercury). Special designs are available for temperatures above 360°C.

Thermometers can, of course, also be calibrated to read in °F.

The clinical thermometer

This is the thermometer for measuring body temperature. Its main features are illustrated in Fig. 7.6 (overleaf).

The range of the *clinical thermometer* needs only to be, say, 95–110°F or 35–43°C.

The small size of the thermometer and rapid entry of heat through the thin-walled bulb allows a reading to be obtained in perhaps as little as half a minute.

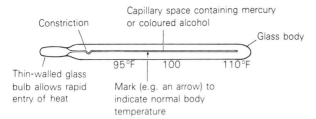

Capillary space containing mercury
or coloured alcohol

Constriction

Glass body

Thin-walled glass
bulb allows rapid
entry of heat

95°F   100   110°F

Mark (e.g. an arrow) to
indicate normal body
temperature

Fig. 7.6   The clinical thermometer

The thermometer may be read after removal from the patient/client because the constriction in the capillary tube prevents return of the mercury to the bulb until the thermometer is shaken in such a way as to force the mercury towards the bulb.

When a mercury thermometer breaks, the glass must be handled with care and all mercury droplets must be collected and not dispersed. Mercury does not vaporise easily but the vapour is poisonous so good ventilation is required wherever mercury has been spilt.

Alcohol-in-glass thermometers

Alcohol can be used in a thermometer, but only up to 78°C (the normal boiling point of alcohol), and the alcohol has to have some colouring added to make it visible. It has the advantages that it can be used for lower temperatures than can the mercury thermometer and it is cheaper.

Maximum-and-minimum thermometers

Thermometers of this type can be reset and then left for a period of time, perhaps a day or a week, and the highest and the lowest temperatures for that period are able to be read. One type of *maximum-and-minimum thermometer* is illustrated in Fig. 7.7. It works in this way. When the temperature rises, the alcohol in the bulb expands and the mercury is pushed round the tube, its level rising in the right-hand tube and carrying the maximum index up. When the temperature falls, the mercury moves so as to fall in the right-hand tube leaving the maximum index unaffected and rising in the left-hand tube, moving the minimum index up. The highest temperature reached is shown by how far up the right-hand tube the maximum index has been left; the minimum temperature is shown by the height to which the minimum index has been taken. The thermometer is reset by pulling each index down with a magnet.

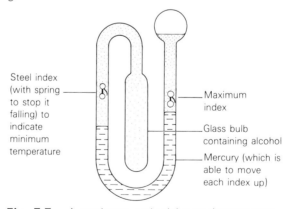

Steel index
(with spring
to stop it
falling) to
indicate
minimum
temperature

Maximum
index

Glass bulb
containing alcohol

Mercury (which is
able to move
each index up)

Fig. 7.7   A maximum-and-minimum thermometer

Bimetal thermometers

Fig. 7.8 shows the principle on which this type of thermometer can work. The spiral is made of bimetal strip. A change of temperature makes the spiral open up or curl up more tightly and the pointer rotates.

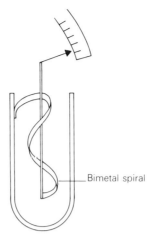

Fig. 7.8   The principle of a bimetal thermometer

Liquid-in-metal thermometers

An example of such a thermometer is the use of mercury in a steel container, and Fig. 7.9 shows how a *liquid-in-metal thermometer* can work.

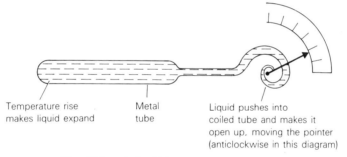

Temperature rise
makes liquid expand

Metal
tube

Liquid pushes into
coiled tube and makes it
open up, moving the pointer
(anticlockwise in this diagram)

Fig. 7.9   A liquid-in-metal thermometer

Other thermometers

Some thermometers make use of expansion of a gas while others depend upon the effect of temperature change on the resistance of a metal or, in the case of *thermistors,* of special materials whose resistance decreases rapidly. when the temperature rises.

Thermostats and temperature controllers

The word *thermostat* best describes a device that maintains one chosen temperature, while a *temperature controller* describes a device for providing a choice of temperature that can easily be varied, usually by use of a temperature selector dial.

Thermal cut-outs are safety devices which switch off a piece of equipment if it becomes abnormally hot due to a fault or to misuse. For example a hair dryer or a fan heater may have a bimetal cut-out that switches off the electricity supply if the air flow is obstructed and overheating results.

A bimetal thermostat/temperature controller

The working of a temperature controller of this type is explained in Fig. 7.10.

Temperature controllers like this are used on some wax heaters used in the beauty clinic.

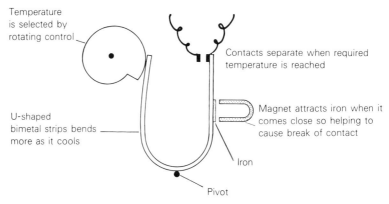

Fig. 7.10   A bimetal thermostat/temperature controller

Liquid or gas expansion thermostat

Thermostats employing mercury in a metal system can be seen in some saunas.

An example of the use of expansion of a gas is shown in Fig. 7.11.

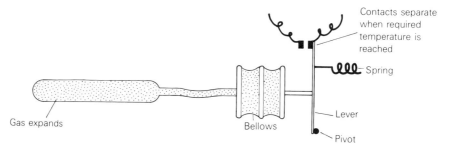

Fig. 7.11   A thermostat making use of the expansion of gas

The electronic thermostat

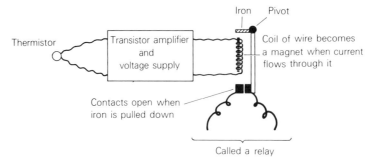

Fig. 7.12   An electronic thermostat system

One possible system is shown in Fig. 7.12. When the temperature of the thermistor is sufficiently high an appreciable current can flow and, after being amplified by the transistor amplifier, this current flows through the relay coil and makes it act as a magnet. The magnet action pulls on the iron part of the lever, and so the lever opens the contacts of a switch. This switch, therefore, stops the flow of electric current to the heater.

For more information regarding amplifiers, see Chapter 13.

The electric kettle plug

If an ordinary electric kettle is allowed to heat up with no water in it, the heating element gets too hot and this makes the plug of the electricity supply cable shoot out from the kettle. This protective action occurs because the movement of a bimetal strip, or the melting of a metal with a low melting point, allows a spring to move so that a rod ejects the plug. The bimetal design allows the rod to be subsequently pushed back, re-setting the spring.

The same plugs are commonly used for the water heaters in steam baths (see Chapter 9).

An electric kettle can also have a bimetal strip which switches off the kettle when the water boils and steam reaches the strip.

HEAT AND HEATING APPLIANCES

Heat

When something is heated the microscopic movements of its atoms or molecules increase. Because greater movement means greater kinetic energy it is clear that energy has to be supplied to cause a temperature rise.

What exactly is heat? To answer this properly two points must be made.

(a) The energy of the microscopic movements of an object's atoms or molecules should be called its *internal energy*, but this is often described as *heat*.

(b) When energy is entering the object from, say, a hot-plate or a flame it is definitely called heat.

Heat is defined as energy that is moving because of a temperature difference. It moves 'of its own choice' from a hotter to a colder place.

For example, in Fig. 7.13 heat flows from the *very* hot wire into the hot-plate. Then this energy passes from the hot-plate into the wax pot and finally from the wax pot into the wax. Once inside the wax the energy goes no farther so that, to be very correct, we should no longer call it heat but the *internal energy* of the wax. Most people will accept the term *heat* instead of internal energy.

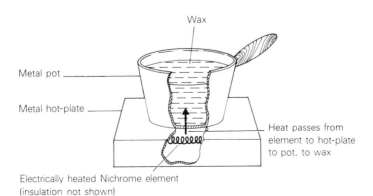

Wax

Metal pot

Metal hot-plate

Heat passes from element to hot-plate to pot, to wax

Electrically heated Nichrome element (insulation not shown)

Fig. 7.13 Heating wax

Specific heat capacity

Just as a sponge or piece of blotting paper needs a lot of water before it seems very wet so some objects need a lot of heat to produce a small temperature rise. Such objects are said to have a large *heat capacity*. We define heat capacity as the number of joules or calories needed for a 1 °C temperature rise.

To compare the heating of different materials such as water, iron, wax, etc., we speak of the *specific heat capacity* (or SHC), meaning the heat capacity per kilogram.

For example, to warm 1 kg of water by 1 °C requires 4200 J, i.e. the specific heat capacity of water is 4200 J per °C per kg.

Clearly, the amount of heat required to produce a particular temperature rise can be calculated by multiplying the specific heat capacity by the number of kilograms and by the number of degrees of temperature rise.

Heat taken in or given out = **Mass × SHC × Temperature change**

The glass of a mercury-in-glass or alcohol-in-glass thermometer should have a small heat capacity so that most of the heat entering it will go into the mercury or alcohol and warm it quickly, little being needed to warm the glass.

Some of the effects of heated masks applied to the face or neck depend upon the thermal capacity of the mask. The stored heat passes from the mask into the skin and causes *erythema*.

Storage heaters must have a large thermal capacity so that a lot of energy can be stored. In Fig. 7.14 the blocks of brick-like material are heated electrically, usually during the night when electricity charges are lower, and then the energy stored in them escapes throughout the day and warms the surrounding air.

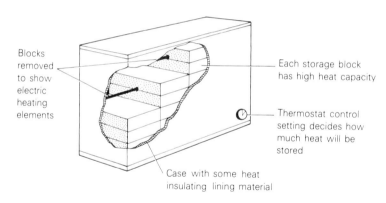

Blocks removed to show electric heating elements

Each storage block has high heat capacity

Thermostat control setting decides how much heat will be stored

Case with some heat insulating lining material

Fig. 7.14   One type of storage heater (a storage radiator)

Heating by gas

Although gas can be obtained in small quantities from cylinders (usually butane gas), most gas heaters are connected to the mains gas supply. In the United Kingdom at present mains gas is largely natural gas obtained from beneath the North Sea. This gas contains hydrogen and methane (CH_4, see Chapter 18) which burn to release large amounts of heat and form carbon dioxide and water which, together with some unburnt gas and incompletely burnt gas (poisonous carbon monoxide particularly), need to be dispersed or taken away via a flue.

In Fig. 7.15 a gas-burning water heater is illustrated. The hot gases resulting from the combustion of the fuel gas pass close to a metal (copper) tube in which water is flowing. The length of tube is sufficient for the water to be hot when it finally emerges. This heater provides instant hot water; a whole tank of water does not have to be heated.

Provided that the pilot flame has been previously lit, the gas burner lights up automatically when the water

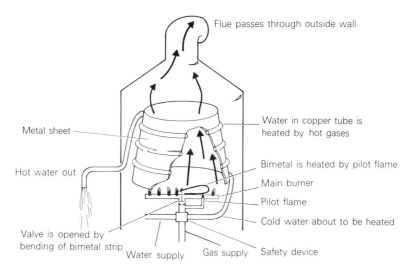

Fig. 7.15 A diagram to show the principles involved in an instantaneous, gas-fuelled, water heater

supply is turned on. If the pilot is not alight then no gas comes from the main burner so that a dangerous build-up of gas cannot occur.

Another example of gas heating is the room heater shown in Fig. 10.17 of Chapter 10.

If a gas leak is suspected in a gas-burning appliance there may be a danger of fire or explosion or of suffocation (*asphyxia*) even if the gas is not poisonous. The main gas tap should be turned off (see Fig. 7.16) and professional help sent for. Windows and doors should be opened. No naked flames should be used, cigarettes should be put out, and no electrical switches should be operated.

If a gas appliance is to be disconnected, the main tap must be turned off first.

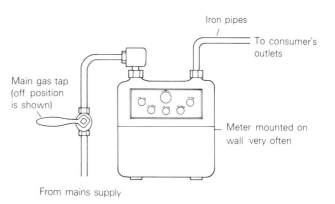

Fig. 7.16 The typical appearance of gas meter and main tap

Your gas bill

For calculations concerning gas supply, energy is often measured in British Thermal Units (Btu) instead of joules or calories. Also the quantity of gas used is often measured in cubic feet rather than cubic metres. The cost for the consumer is decided by the number of joules or Btu of energy which can be obtained by burning the gas. As seen in Fig. 7.17 (b), the gas supply company gives the number of Btu obtainable per cubic foot of gas (the calorific value). The way in which the cost is worked out, from the cubic feet used and the calorific value, should be clear from the gas bill illustrated in Fig. 7.17 (b), overleaf.

To read a gas meter such as the one shown in Fig. 7.17 (a): if there is a figure exactly at the pointer this is recorded, otherwise the figure recorded is the lower one. This should be clear from the illustration.

(a)
This pointer moves fastest and is for test purposes
Each div. 100 000
Each div. 10 000
Each div. 1 000
Each div. 100 cu ft
Each div. 10 cu ft. (100 per rev)
The reading is:-
759360 cu ft
OR
700 000
+ 50 000
+ 9 000
+ 300
+ 60
759 360

(b)

SOUTH WEST GAS
(A part of The British Gas Corporation)

P.O. BOX NO. 31, BATH, BA2 4EP.

VAT REGISTRATION NUMBER 232 1770 91

REFERENCE FOR ENQUIRIES	TARIFF	CALORIFIC VALUE	DATE OF READING	DATE OF ACCOUNT (TAX POINT)
5153204 2	DOMESTIC	38.3 MJ/m³	6 SEP 82	7 SEP 82

MR A C BRIDGE
17 WHEATSTONE ROAD
GLOUCESTER
GL3 9ZP

Please refer enquiries to SOUTH WEST GAS AT ABOVE ADDRESS
TEL BATH 61141 (GLOUCESTER GAS SECTION)

METER READINGS: PRESENT	PREVIOUS (E - ESTIMATED, S - SELF READ)	GAS SUPPLIED: CUBIC FEET (HUNDREDS)	THERMS	PENCE PER THERM	CHARGES £	VAT CODE	VAT CHARGES £	TOTAL AMOUNT £
8532	8466	66	67.716	30.500	20.65	A	0.00	20.65
STANDING CHARGE					9.00	A	0.00	9.00

VAT CODE A VALUE £29.65 @ ZERO % £0.00

TOTAL DUE (INCLUDING VAT) 29.65

(1026 Btu's per cu ft)

In hundreds of cubic feet
1 therm = 100 000 Btu
contains 6600 × 1026 Btu
= 6771600 Btu
calorific value
(equals 38.3 MJ/m³)

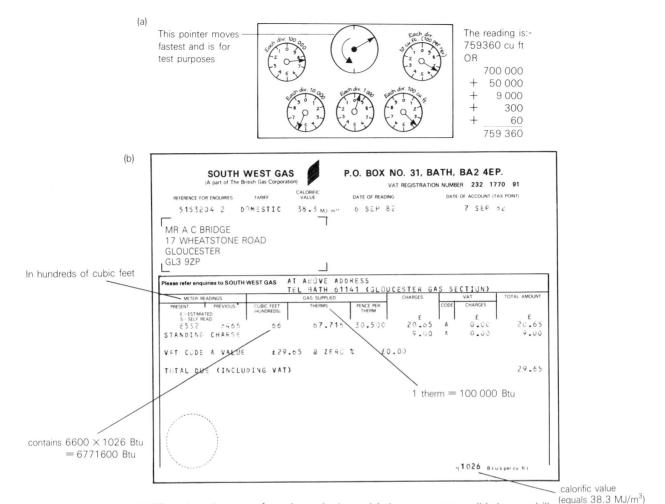

Fig. 7.17 How the cost of gas is worked out: (a) the gas meter; (b) the gas bill

Electric heating in the beauty salon

Uses include:

- General heating of the salon using perhaps storage radiators (Fig. 7.14, p. 70) for background heating plus fan heaters (see Chapter 10).
- Saunas, steam baths and showers.
- Water heating for washing hands (and for making tea!)
- Face steamers.
- Wax heaters.
- Hand-dryers and hair-dryers.
- Epilation by diathermy (see Chapter 13).

The electricity bill

It has already been explained in Chapter 6 that the amount of electrical energy used by the consumer is usually stated in kilowatt-hours, also called 'units' (the term 'unit' here meaning the UK Board of Trade Unit of electrical energy).

The appearance of an electricity bill is shown in Fig. 7.18.

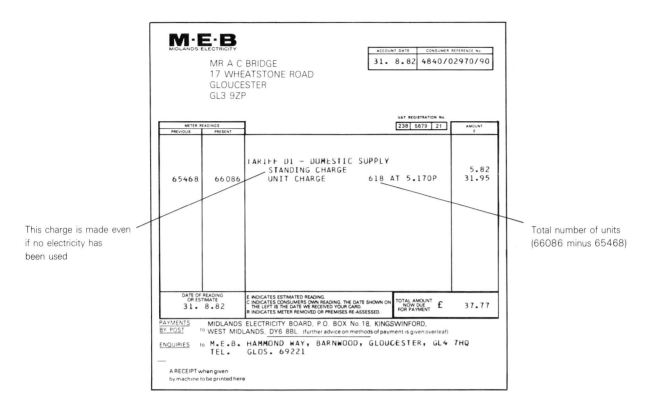

This charge is made even if no electricity has been used

Total number of units (66086 minus 65468)

Fig. 7.18 An electricity bill

Heating by friction

If a surface is pressing against another and slides along it, then this rubbing, or *friction*, movement produces some heating, more so if the surfaces are rough.*

Advantage can be taken of this effect to warm the hands by gently rubbing them together. But friction can also be a problem when the moving parts of a motor or other machine rub so much as to produce excessive heating.

Vigorous rubbing of the skin, or *massage* (Fig. 7.19), not only creates warmth. It must also cause some microscopic damage to cells so that a chemical called *histamine* is released. This in turn causes *erythema* (redness and warming of the skin). In the massage technique known as *effleurage*, stroking movements are employed so that the hands rub over the skin surface. *Pétrissage* is the friction method where the fingers

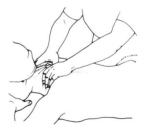

Fig. 7.19 Massage using friction

or hands stay on one area of skin but make the flesh move over underlying bone or muscle. Even in the kneading method of pétrissage and in *tapotement* (percussion movements) some rubbing must occur within the tissues as they are moved.

---

*This is because microscopic bumps, that are present on even the smoothest surfaces, must push each other out of the way or else some must break off. In the process, atoms in the surfaces are made to move and the surfaces become heated.

SUMMARY

- Temperature is hotness or coldness measured in a reliable way.

- Temperature is measured with a thermometer.

- Using the Celsius scale, freezing water is 0°C, boiling water is 100°C.

- Using the Fahrenheit scale, freezing water is 32°F, boiling water is 212°F.

  Normal body temperature is 98.4°F or 37°C.

- $°C = \left(°F - 32\right) \times \dfrac{5}{9}$ $°F = \left(°C \times \dfrac{9}{5}\right) + 32$

- Heating means raising the temperature and causes atoms and molecules to increase their microscopic movements, such as vibrations.

- Thermal expansion is the increase of size that nearly always accompanies temperature rise. 'Things expand when heated and contract when cooled.'

- Uneven expansion can cause ordinary glass to crack if very hot water is poured on it.

- Expansion of mercury is made use of in mercury-in-glass thermometers.

- Unlike a general purpose thermometer, a clinical thermometer has a constriction to stop the mercury returning before the reading is taken.

- A bimetal strip bends when it is heated.

- Bimetal strips can be used for
  ○ thermometers,
  ○ and for temperature control, e.g. of room heaters and wax heaters.

- An object of large heat capacity needs a lot of heat to produce a small temperature rise.

- The energy provided by gas used, in the United Kingdom, is measured in Btu instead of joules. 100 000 Btu is 1 therm.

- Rubbing is called 'friction' and produces heat.

EXERCISE 7

1. Suggest values for the following temperatures:
(a) the normal boiling point of water
(b) water that is freezing
(c) ice taken from a refrigerator
(d) salon air temperature.

2. If the temperature recorded in a steam bath is 45°C, what is its value in °F?

3. The melting point of some paraffin wax is advertised as 122°F. Convert this to °C.

4. If the temperature of a sauna is found to be 194°F, what is this in °C?

5. The temperature of hot water supplied by a hot water system is recommended to be 60°–70°C (60°C being preferred if the water is hard to prevent scaling of heaters). Convert these temperatures to °F.

6. Give one example of thermal expansion being a nuisance or a hazard.

7. How is a simple bimetal strip constructed? Explain one of its uses.

8. Give one advantage that an alcohol thermometer has compared with a mercury thermometer.

9. Give two differences between a clinical thermometer and a general purpose mercury-in-glass thermometer.

10. Draw a labelled diagram of a mercury-in-glass thermometer.

11. Suggest how a maximum-and-minimum thermometer could be useful in a beauty salon.

12. Mercury-in-copper thermostats are used in some saunas. Why is mercury suitable?

13. What is a thermostat? (CGLI)

14. Explain how an electric kettle can be made to switch off automatically when the water boils.

15. When a metal hot-plate is warmed up, what is happening to the atoms of which it is made?

16. Name three different units suitable for measuring heat.

17. What is a storage radiator?

18. Gas is leaking from a pipe leading to a water heater. What action would you take?

19. You smell gas from a water heater and notice that the pilot flame has disappeared, probably blown out because of a wide-open window nearby. What action would you take?

20. Sketch the positions of the pointers on a gas meter that is reading 305740 cubic feet. (You will need to refer to Fig. 7.17 (a).)

21. With reference to Fig. 7.17 (b); how many Btu would have been used if the volume of gas used was 10 000 cubic feet.

22. If 1 Btu is enough heat to warm 1 cu ft of water by 1°F, how many Btu are required:

(a) to heat a $2 \times 2 \times 2$ ft tank of water from 10°C (50°F) to 60°C (140°F)?

(b) to heat a bath of 5 cu ft of water from 10°C (50°F) to 40°C (104°F)?

What is the cost in each case if the tariff is 20p per 100 000 Btu for the gas supplied?

23. What is meant by 'standing charge'?

24. Explain what is meant by: (a) heat capacity, (b) specific heat capacity.

25. What is the approximate temperature of application for the following treatments: (a) paraffin wax, (b) steam bath, (c) sauna bath? (CGLI)

8

AIR PRESSURE AND VACUUM

AIR PRESSURE

The main purpose of this Chapter is to explain the working of vacuum massage equipment, but fans and pumps are included.

The pressure of a gas

A gas pushes upon any surface in contact with it because the countless molecules of the gas are for ever moving and hitting the surface. The gas is producing a pressure on the surface. This is the pressure of the gas or the *gas pressure*.

When a gas is squeezed its volume decreases only until the gas pushes back with a pressure equal to the pressure applied as shown in Fig. 8.1. For this reason the pressure of a gas is telling us the pressure applied to it.

The pressure *of* a gas means the pressure exerted *by* the gas and equally the pressure acting *on* the gas.

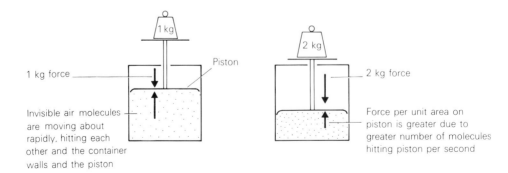

Fig. 8.1 A gas, e.g. air, under pressure

The effect of pressure on density of a gas

As Fig. 8.1 shows, the molecules occupy a smaller volume so that the gas density is greater when the pressure is greater. In fact, the pressure of a gas can be a useful indication of its density.

The effect of temperature on density of a gas

When a gas is warmed, its molecules move faster, its pressure rises, and so it 'pushes out', i.e. *expands*, unless the container prevents this. This effect is used in gas thermometers and temperature controllers (see Chapter 7). Expansion of a gas causes its density to be reduced.

Atmospheric pressure

It has already been explained in Chapter 2 that the air pushes down on objects beneath it and, we might add, on other air beneath it.*

Gas flow

In Chapter 2 it was stated that *gas flow* requires a pressure difference. It is most likely that the pressure difference is caused by a pump of some kind. Otherwise it may be caused, for example, by natural convection in which some air has been heated and has therefore expanded and become less dense. It consequently rises through the heavier, cooler air.

Air pumps

A good example of a pump is the working of the lungs. To breathe in we make our lungs increase in volume. This reduces the air pressure in the lungs so that air enters from the higher (atmospheric) pressure outside (Fig. 8.2).

When the lungs are allowed to relax to their normal size, the air is squeezed in the lungs so that it flows out to the lower atmospheric pressure outside.

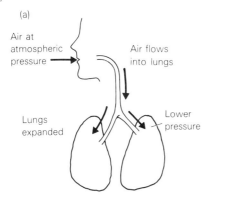
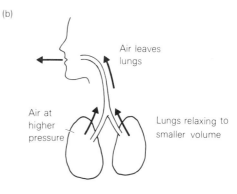

(a)
Air at atmospheric pressure
Air flows into lungs
Lungs expanded
Lower pressure

(b)
Air leaves lungs
Air at higher pressure
Lungs relaxing to smaller volume

Fig. 8.2 How air is made to enter and leave the lungs: (a) breathing in; (b) breathing out

*This effect is, of course, due to gravity pulling down on every molecule of air while each molecule of air is supported by being struck on the underside by molecules below.

An example of a simple, electrically driven air pump that is used in some beauty therapy equipment is the *diaphragm pump* shown in Fig. 8.3.

As the motor shaft rotates, the diaphragm is moved up and down, alternately increasing the chamber volume so that air enters through the one-way inlet valve and, as the diaphragm is pushed down, pushing air out through the one-way outlet valve.

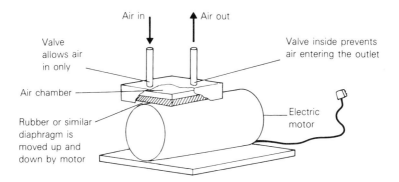

Fig. 8.3   A diaphragm pump

Fans

Fans are really air pumps but their function is to cause air flow rather than to produce significant pressure changes. They are particularly important for room ventilation (see Chapter 10), and two examples are shown in Fig. 8.4.

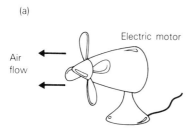
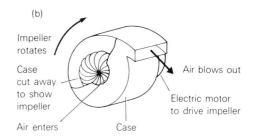

Fig. 8.4   Two kinds of fan: (a) an axial fan;   (b) a centrifugal fan or pump

Vacuum

When most of the air is removed from a container the space inside is called a *vacuum*. The air pressure is then very low. It is impossible to remove every bit of gas from a container so that an exactly zero pressure is not obtainable. A high vacuum means a very low pressure, say 1 millionth of 1 mm of mercury pressure (0.000 13 N/m^2 or pascal).

Removing only a little of the air from a container produces a reduced pressure (i.e. lower than atmospheric pressure) which is only a low vacuum or *partial* vacuum. For vacuum massage a low vacuum of 0–30 cm of mercury (i.e. 0–30 cm of mercury below atmospheric pressure) is commonly used. This is discussed on the following page.

Suction

Suction is the word used for the pull that appears to be caused by a low pressure or vacuum. Often it is accompanied by a noticeable air flow. However, it should be realised that a low pressure or vacuum does not pull on anything. The air can only *push* on things (the push being due to air molecules hitting the thing's surface). When, for example, a liquid is sucked through a drinking straw the rise of liquid is due to a partial vacuum being created in the mouth so that there is little or no air pressure pushing down the straw while the atmospheric pressure outside pushes the drink up. (See Fig. 8.5.)

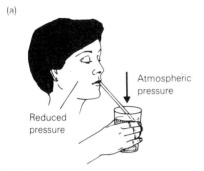

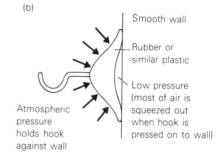

Fig. 8.5   Suction is due to atmospheric pressure: (a) drinking with a straw;   (b) a suction hook

Implosion

If a vacuum is present inside a container, then the strength of the container must be sufficient to withstand the atmospheric pressure which pushes on the outside. If the container is not strong enough it will collapse inwards, probably suddenly and violently. This *implosion* of a glass container can be very dangerous because fast-moving fragments of glass can hit someone nearby.

Implosion is most likely when the container is large or insufficiently thick or of a shape that is weak in places. For these reasons only apparatus recommended for vacuum work should ever be evacuated.

PRESSURE MEASUREMENTS

Manometers

To measure how high or low a pressure or vacuum is we use pressure and vacuum gauges and *manometers.*

The words *gauge* and *meter* each mean *measuring instrument,* but it is customary to use the word 'gauge' for gas pressure measurements. The term *manometer* means *pressure-measuring instrument* and is always used for U-tube measurement of pressure.

Simple U-tube manometers

The simplest *U-tube manometer* consists of a U-shaped glass tube connected on one side to the apparatus whose gas pressure is to be measured, as seen in Fig. 8.6. (overleaf).

The open U-tube manometer is suitable for testing pressures not too much higher or lower than atmospheric pressure. The height (*h*) is measured and the pressure is described as '*h* cm or mm of water' (or mercury or

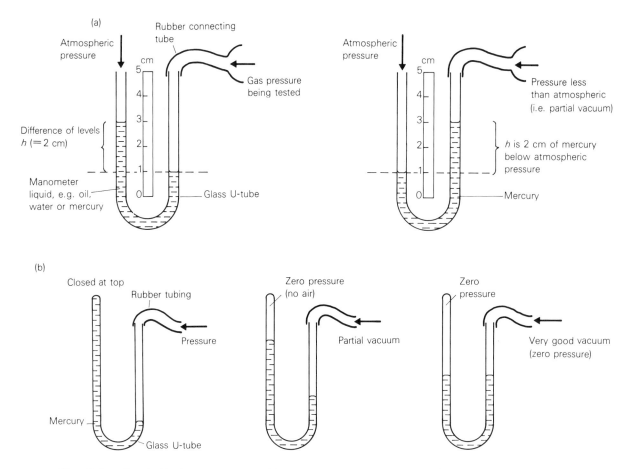

Fig. 8.6 Simple U-tube manometers: (a) open U-tube manometer;   (b) closed U-tube manometer

whatever liquid) above or below atmospheric pressure, as shown in Fig. 8.6 (a). For example:

2 cm of water vacuum implies 2 cm of water below atmospheric pressure.

A denser liquid like mercury will give a smaller difference in levels, i.e. a smaller reading h, and is preferred if the pressure to be measured is not very close to atmospheric pressure. A less dense liquid, i.e. a lighter liquid, is preferred if the pressure is not very different from atmospheric.*

In Fig. 8.6 (b) is shown a closed U-tube manometer in which there is no air in the closed side of the tube. It is suitable for testing a fairly high vacuum, greater than is likely in vacuum massage.

The Bourdon gauge

This is a gauge which may be found as part of a vacuum massage apparatus. It works on the same principle as the party toy (Fig. 8.7 (a)) which unrolls when it is blown.

In Fig. 8.7 (b) the pointer moves to the right over the scale when air is taken from the gauge and moves to the left when the pressure in the gauge is increased. If, when

---

*If the pressure being tested is described as P pascal and atmospheric pressure is P_a pascal and the area of the cross-section of the tube is A metre^2 then the force down on one side of the tube is $P \times A$ and on the other side is $P_a \times A$ newton. This difference makes the liquid move round the tube until the difference $PA - P_aA$ is equalled by the weight of extra liquid on the higher side, which is $hA\rho$ kg force or $10hA\rho$ newton, where ρ is the density of the liquid. Thus $hA\rho = \dfrac{PA - P_aA}{10}$ or $P - P_a = 10\rho h$, i.e. P_a in pascal (Pa) is given by $10\rho h$ where ρ is density in kg/m^3 and h is the manometer reading in metres.

the pressure is atmospheric, the pointer is near the middle of the scale then pressures above and pressures below atmospheric can be indicated. Otherwise the gauge may work for pressure only or for vacuum only.

By including cogs and levers the pointer can be made to move quite a lot when the pressure changes only a little, i.e. the gauge can be made sensitive.

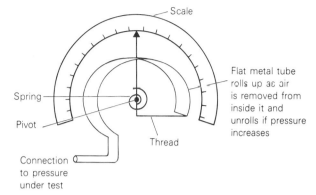

Fig. 8.7   The Bourdon gauge: (a) the principle on which it works;   (b) a simple design of Bourdon gauge

VACUUM MASSAGE

Vacuum massage equipment

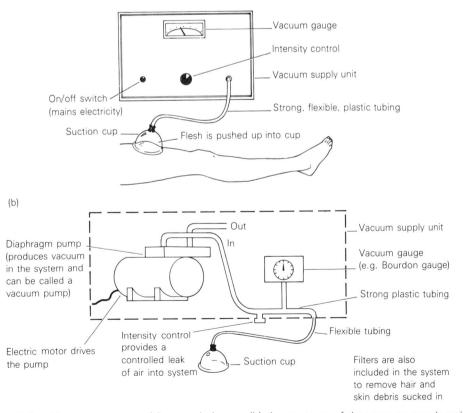

Fig. 8.8   Vacuum massage: (a) general view;   (b) the contents of the vacuum supply unit

In this massage technique a partial vacuum is produced under a cup placed against the skin. The cup is transparent, usually being made of a clear plastic, so that the skin and underlying flesh can be seen to be sucked into the cup (or rather they are pushed into the cup by the atmospheric pressure acting all around and within the body). A general view of the apparatus is shown in Fig. 8.8 (a) and Fig. 8.8 (b), and shows the important parts contained within the supply unit. The pump, usually of the diaphragm type, removes air from the system (i.e. from the tubing, cup and gauge) and vents it to the atmosphere through an outlet usually at the back of the supply unit.

The intensity of the treatment is, of course, decided by the vacuum produced. This is varied by the intensity control which is an air inlet tap. When a lot of air is allowed to enter here the pump can produce only a very low vacuum. With the intensity control set for maximum intensity it will be completely closed so that no air leaks into the system and the pump produces its lowest possible pressure of perhaps 300 mm of mercury (6 pounds force per square inch) below atmospheric pressure (300 mm of mercury vacuum).

Suction cups can be released from the skin by easing a finger beneath the rim of the suction cup to allow entry of air, without having to change the intensity control.

Uses of vacuum massage equipment

The movement of the surface body tissues under the suction cup causes *erythema* (redness due to increased blood supply) in the tissues and promotes movement of blood through the vessels. If, in addition, the suction cup is made to slide over the previously oiled skin and directed so as to follow the direction of lymph flow in the lymph vessels, then the removal of waste via the lymph system is assisted. So we have vacuum massage for stimulation of the tissues and for increasing lymphatic drainage.

Also, by use of *ventouses* of suitable size and shape the drawing out of *comedones* (blackheads) can be achieved.

An alternative method of vacuum massage uses a static cup with pulsating vacuum. The cup is under vacuum at all times but the vacuum can be automatically increased and decreased alternately, i.e. it is *pulsed*, and the body tissue is forced into the cup strongly then less strongly. Both the minimum and maximum vacuum are adjustable as well as the duration of the pulses. It is usual for the pulsed vacuum supply unit to provide several cups which can cover the body over most of the area requiring treatment (Fig. 8.9 (a)).

Finally, if a cup or ventouse is held in place by hand, alternate vacuum and pressure can be used to provide tapping massage (*tapotement*).

(a)

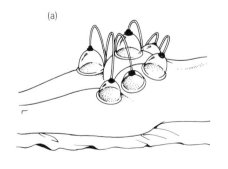

(b)

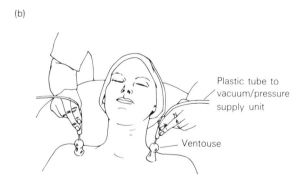

Plastic tube to vacuum/pressure supply unit

Ventouse

Fig. 8.9 Pulsed vacuum: (a) pulsed vacuum using several suction cups; (b) using alternating vacuum and compressed air massage

SUMMARY

- Increasing the pressure of a gas decreases its volume and increases its density.

- The pressure of a gas means the pressure on it or the equal pressure with which it pushes back.

- Expansion of a gas when heated, and its contraction as it cools, can be used in gas thermometers and temperature controllers.

- Air pumps can be of the diaphragm type or may be fans of axial or centrifugal design. (There are other kinds of pump too.)

- Suction works by the pushing of atmospheric pressure.

- Manometers are used to measure pressure differences.

- A Bourdon gauge is a dial gauge (or meter) for measuring pressure or vacuum.

- The main parts of a simple vacuum massage machine are a diaphragm pump, a Bourdon vacuum gauge, a suction cup and an adjustable air inlet.

EXERCISE 8

1. Explain what is meant by *vacuum* and by *suction*.

2. Name the essential parts of a small vacuum massage apparatus.

3. Vacuum massage used incorrectly can cause discomfort and even bruising of the client. Name another hazard arising from the improper use of vacuum.

4. Name two benefits possible with vacuum treatment.

5. Suggest a use for an axial fan in the salon. What safety precautions are required in its use?

6. What is a manometer?

7. What is a barometer? (See Chapter 2.)

8. Suggest how an open U-tube manometer could be fitted to a vacuum massage unit and the maximum vacuum then measured in centimetres of mercury.

9. When using a U-tube manometer what would be the effect of changing to a denser liquid?

10. An open, U-tube manometer connected to the mains gas supply contains mercury and shows a difference of levels of 2 cm. Describe the gas supply pressure in N/m^2 (i.e. in pascal, Pa). The density of mercury is 13 600 kg/m^3.

11. A typical vacuum massage unit could produce a vacuum as high as 30 cm of mercury below atmospheric pressure. Given that water is approximately 14 times less dense than mercury, what reading would a water-filled, open, U-tube manometer give?

12. Why is thin, soft plastic tubing not suitable for vacuum work?

9

LIQUIDS, MELTING AND EVAPORATION

INTRODUCTION

Entering a poorly designed salon on a winter's day you could be faced with the problems of water frozen in the supply pipes, water dripping from the walls because of evaporation and condensation, clouds of steam in the shower room and puddles of water on the floor. Our interest in liquids, melting and evaporation would then be obvious. In the salon we are also interested in the properties of cosmetic liquids and massage oils, in the melting of waxes used for hair removal, and in the evaporation of perfume and other cosmetic ingredients.

CHANGES OF STATE

The three states of matter

The three conditions or states in which a substance can exist are *solid, liquid* and finally *gas* or *vapour*.* The differences between them can be explained by the way in which their molecules are arranged.[†]

The change of state from solid to liquid is, of course, called melting and evaporation is the conversion of liquid to vapour. The names given to other changes are shown in Fig. 9.1.

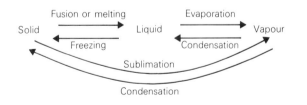

Fig. 9.1   Changes of state

---

*For the moment we will speak of vapour rather than gas and explain the exact difference between these terms later.

[†] As explained in Chapter 3, a solid has its atoms or molecules close together, held in place by bonds between them. The intertwining of long molecules as in plastics, waxes and glasses also holds the molecules in place. In crystalline solids like frozen water the molecules also are tidily arranged, that is to say, ice has a regular molecular structure.

In the liquid state flow is possible because there are gaps in the structure and molecules have become free of each other to some extent.

In the vapour state molecules are well separated so that attractions between them are negligible.

The structures of ice, liquid water and water vapour are illustrated in Figs 3.10 (b), 3.11 and 3.12 on pp. 20–1.

– The effects of temperature and pressure on the state of a substance –

The atmospheric or other pressure on a substance tends to squeeze the atoms or molecules together and so encourage the solid state and discourage the vapour state. Raising the temperature of a substance makes its atoms or molecules move more vigorously and so encourages the particles to shake loose from each other and to take up more space. Thus higher temperatures favour the vapour state and lower temperatures produce the solid state.

Melting and freezing*

In the case of water and other substances whose solid state is crystalline, melting occurs at a definite temperature called the *melting point* (MP). For water the melting point is 0°C or 32°F (assuming normal atmospheric pressure of about 76 cm of mercury). If liquid water is cooled, then it freezes at this temperature. Thus the melting point and *freezing point* are effectively the same.[†]

Any attempt to raise a solid's temperature above its melting point by supplying heat rapidly simply quickens the melting.

Fig. 9.2 is a graph showing the effect of supplying heat at a steady rate to some ice after removing it from a refrigerator.[‡]

The heat taken in during melting and given out during freezing is called *latent heat* (or 'hidden heat', because it seems to hide in the liquid rather than show its presence by producing a temperature rise).

The specific latent heat of fusion, meaning the amount per kilogram, is approximately 3300 J per kg for water.

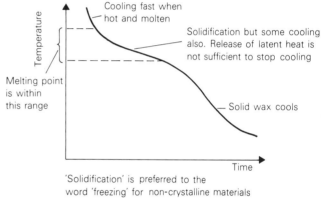

'Solidification' is preferred to the word 'freezing' for non-crystalline materials

Fig. 9.3   The solidification, or freezing, of a wax

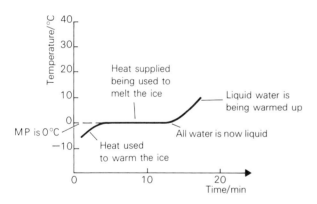

Fig. 9.2   The steady heating of water, initially frozen

Because many waxes are not crystalline they may melt over a range of temperatures as shown in Fig. 9.3. A paraffin wax may melt over a range of only two degrees from 50°C to 52°C, but other waxes may have less well defined melting points.

---

*In the salon the wax or water whose state we may wish to change will be, like most things, acted on by atmospheric pressure, no more or less, and melting or freezing will be achieved by warming or cooling the substance.

[†] At the melting point the attractive forces (bonding forces) are no longer sufficient to prevent the vibrations of the particles from shaking the solid structure apart.

Energy is used to push molecules apart when melting occurs so that melting would produce cooling and the melting would immediately stop if heat were not continuously supplied.

[‡] If some liquid water has heat steadily taken from it, it cools to 0°C, then freezes and then cools below 0°C, following the graph of Fig. 9.2 but beginning on the right of the graph and proceeding towards the left. During freezing, the energy of separation of the molecules becomes heat so that heat is taken from the water without fall of temperature.

Another paraffin wax of slightly different chemical composition might have an MP as low as 35°C or as high as 75°C.

Carnauba wax, an important ingredient in lipsticks, melts between 80 and 87°C. Beeswax which is used in many cosmetics, including lipsticks and depilatories, melts between 62 and 64°C.

Change of volume during melting and freezing

It is to be expected that when a solid melts, its volume should increase and its density decrease as a result of its molecules having separated to some extent. Waxes show this well when we look at the surface of some wax that has solidified in a pot (Fig. 9.4 (a)). Applied as a face mask, wax contracts as it solidifies and provides a suitable feeling of slight tightening.

The volume change that occurs when water freezes is unusual because it is an expansion. This is because ice has a very open structure as seen in Fig. 3.10 (b). Expansion of water as it freezes commonly results in burst water pipes in very cold weather, lead pipes splitting as seen in Fig. 9.4 (b). It is when the temperature rises after the freeze that the ice melts and the burst becomes evident, with water pouring from the pipe.

(a)

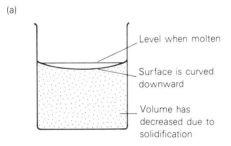

Level when molten

Surface is curved downward

Volume has decreased due to solidification

(b)

Ice

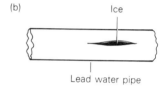

Lead water pipe

Fig. 9.4   Volume changes due to freezing (solidification): (a) wax contracts when it solidifies;   (b) a burst water pipe

Freezing of salt water

The presence of common salt in water has the effect of making freezing more difficult, meaning that the freezing point is lower. For this reason salt can be spread on footpaths to avoid a slippery coating of ice being formed by frozen rainwater or snow.

Evaporation

Fig. 9.5 shows the difference between *ordinary evaporation* and *boiling*.

Ordinary evaporation means the escape of vapour from only the surface of a liquid. It occurs at all temperatures but is greater at higher temperatures.

Boiling is the escape of vapour, usually very rapid, from all parts of the liquid, not only from the surface but by bubbles of vapour from within the liquid. Boiling occurs at a particular temperature called the boiling point (BP). The boiling point of water is 100°C (212°F) if the pressure on the water is just normal atmospheric pressure.

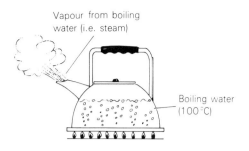

Fig. 9.5 Ordinary evaporation and boiling

The cause of ordinary evaporation

(The molecular explanation is given in the footnote, but this section can be omitted.*)

The rate of loss of liquid by ordinary evaporation†

Warming the liquid makes more vapour leave the liquid, increasing the net loss.

Wind, or any other draught of air, takes vapour away so that it cannot return to the liquid and the net rate of evaporation is therefore increased (Fig. 9.6).

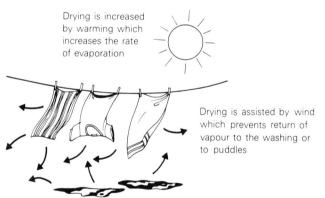

Drying is increased by warming which increases the rate of evaporation

Drying is assisted by wind which prevents return of vapour to the washing or to puddles

Fig. 9.6 Drying by ordinary evaporation

Evaporation of liquid into an enclosed space, for example, in a sealed bottle (see Fig. 9.7, overleaf), will finally result in the space becoming so filled with vapour that as many molecules return to the liquid every second as are leaving the liquid. The net rate of evaporation is then zero. The space is said to be *saturated*. The vapour has reached its maximum concentration and it too is described as saturated. It is usual to describe the saturated vapour not by its concentration but by the pressure it exerts. This *saturation vapour pressure* is denoted by SVP.

For example, water evaporating into an air-filled space at 22°C produces saturation when the vapour density is 20 g per m^3 and the saturation vapour pressure is 2620 N per m^2:‡

The saturation density and SVP increase with temperature — the amount of water vapour that can be held in the air increases with temperature.

---

*Within a liquid or solid the molecules are moving, some faster than others. If a molecule happens to be hit rather a lot on one side by other moving molecules it will be made to move very fast and become free and it may shoot out of the liquid surface. Outside of the surface such molecules are free and well separated and so are in the vapour state. Evaporation has occurred. The same effect in a solid is responsible for sublimation which is explained on p. 91.

Evaporation in this way is more likely to happen if the temperature of the liquid is higher.

†Molecules that have left a liquid mix with the air or vapour already outside and may return to the liquid. The rate of loss of liquid, or in other words, the net rate of evaporation, equals the difference between vapour leaving and vapour returning.

‡If the temperature within the enclosed space is raised, then the evaporation from the warmer liquid is more rapid and a higher concentration and vapour pressure are reached before the return of molecules equals the increased rate of evaporation.

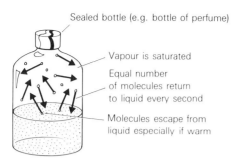

Fig. 9.7 Saturation

Volatile liquids

A liquid which has a strong tendency to evaporate is described as a *volatile liquid*. (Weak forces between the molecules of liquid cause this.)

Perfume *needs* to be volatile. Evaporation of the perfume enables vapour to reach the nose. However, if it is too volatile it evaporates and is all gone too soon.

A volatile liquid will produce a high concentration of vapour in an enclosed space and so it has a high SVP.

Latent heat of evaporation

Energy was needed to cause melting and again energy is needed to produce evaporation because here too work has to be done to separate molecules.

The energy needed to evaporate a kg of liquid is called the specific *latent heat of vaporisation*. When vapour condenses the latent heat is given out again.*

Cooling by evaporation

Evaporation, especially if helped by a draught and especially if the liquid is very volatile, can produce very noticeable cooling.

This can be demonstrated as suggested in Fig. 9.8.

Intense cooling (loosely described as 'freezing'), to anaesthetise the skin before ear-piercing or for a treatment such as certain wart-removal operations, can be obtained by spraying a very volatile liquid on to the skin.

Domestic refrigerators work by evaporating a liquid in pipes where the cooling is required. The vapour is then liquefied again in another part of the refrigerator where the latent heat released can be lost into the room.

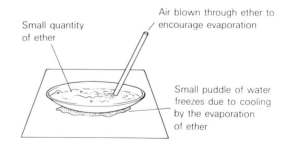

Fig. 9.8 A demonstration of cooling by evaporation carried out in the open air or a laboratory fume cupboard.

---

*Ordinary evaporation can occur without heating and in this case the latent heat needed is obtained at the expense of the energy of molecule movements in the liquid so that cooling occurs. Looking at this in another way, we realise that it is the faster molecules that leave the liquid so that the remaining liquid consists of slower moving molecules. It has been cooled.

The cooling action of cold cream applied to the skin largely depends on the evaporation of water from the cream.

Sweating from the body makes use of cooling by evaporation to prevent overheating of the body.

The evaporation of water from a damp cloth can provide useful cooling of perhaps some food or drink and can reduce bruising of the skin after a blow, but cooling of damp clothes can chill the body and lead to one's catching a cold.

Evaporation from calamine lotion cools and thereby helps to sooth the skin when applied, for example, after sunburn.

Explanation of boiling

(The molecular explanation is given in the footnote, but this section can be omitted.*)

Latent heat and boiling

When a liquid is heated to its boiling point, the escape of vapour is rapid enough to prevent further temperature rise and all the heat supplied becomes latent heat of vaporisation. The graph in Fig. 9.9 shows the effect of steady heat supply to some water which finally becomes steam.

When steam condenses the latent heat is released again. Because of the high specific latent heat of steam, a small amount of steam gives out a lot of heat when it condenses. Consequently, scalds by steam can be quite serious burns and steam condensed in the air inside a steam bath or Turkish bath provides a lot of heat.

Steam is invisible. Visible 'steam' consists of minute droplets of hot water formed by steam mixing with air and condensing.

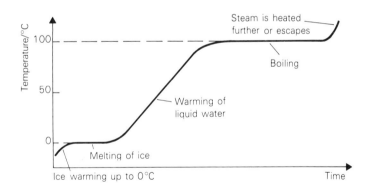

Fig. 9.9 Heating water to produce steam

---

*In a liquid, especially if it is hot, microscopic bubbles of vapour must occur wherever a number of fast-moving molecules happen to meet and push each other apart. But these bubbles are immediately squashed by the surrounding liquid upon which atmospheric pressure is pushing so that the vapour is squeezed back to the liquid state. If now the temperature of the liquid is increased, a temperature is reached where the (saturated) vapour pressure in the bubbles equals the atmospheric pressure. The bubble pushes out equal to the atmospheric pressure and survives. As more bubbles are produced they join and rise as bubbles of increasing size and escape through the liquid surface. The temperature at which this boiling occurs is, of course, the boiling point. Any attempt to raise the temperature above the boiling point by more rapid heating results only in a minute temperature rise leading to increased evaporation and cooling back to the boiling point. Thus greater heating produces more evaporation but no significant increase of temperature.

Face steamers

A hot or warm, moist spray can be applied to the skin, often on the face or the back, using a steamer in which water is boiled and the steam emerges from a pipe, mixes with air and is directed towards the client. A free-standing apparatus is shown in Fig. 9.10. It is very common to include in the apparatus the facility for producing ozone which is added to the steam because of its antibacterial action and influence upon the acidity of the skin. Such a steamer is described in Chapter 17.

Alternatively a steamer may be designed so that various lotions can be included in the steam.

The mixing of the steam with air makes it condense into small droplets of water. So the steam becomes visible and its temperature largely depends on the amount of air with which it mixes, and varies with the design of the steamer.

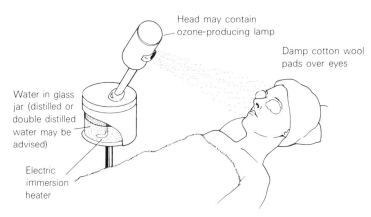

Fig. 9.10   Using a face steamer

Steam baths

A typical steam bath is shown in Fig. 9.11.

The water tank is filled to the recommended level, usually 2 or 3 inches from the top. A towel is used to cover the top opening, and towelling may have to be placed to protect the client from direct, uncooled steam from the tank. The electricity is switched on. If at any time the water level falls below the heater element (or

elements), the cut-out will operate or, in other designs the electric plug may jump out as it does from some electric kettles. A typical working temperature is 40°C. When the required temperature has been reached the client enters and a towel is placed around the neck, as in the diagram.

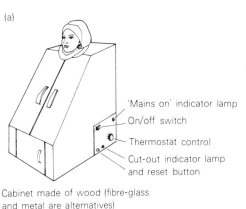

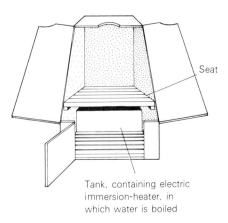

Fig. 9.11   A typical steam cabinet: (a) cabinet in use;   (b) view of interior

The effect of pressure on boiling point

If the pressure on a liquid is increased, a higher temperature will be needed to cause boiling.

Use is made of this fact in the domestic pressure-cooker and the autoclave. In both of these water is heated and boils, but the vapour boiled off builds up in the closed space above the water. As a result the pressure in the apparatus becomes considerably greater than atmospheric pressure and the boiling point of the water increases. The increase is allowed to continue up to one atmosphere above normal atmospheric pressure (i.e. the pressure has doubled), giving a boiling point of about 120°C.

The construction of a small autoclave can be seen in Fig. 22.7 (p. 270).

The reason the increased boiling point is needed is, in the case of the pressure-cooker, that food cooks more quickly at the higher temperature. In the case of the autoclave, which may be used in a salon for sterilising equipment, it is the high temperature that enables the autoclave to kill micro-organisms (germs) so effectively (see also Chapter 22).

Vapour or gas

High temperatures encourage the vapour or gas, rather than the liquid or solid, state and for every substance there is a temperature above which it just cannot be a liquid no matter how much it is 'squeezed'. For water this temperature is 374°C. Above this temperature water is a gas. Below this temperature we can use the term *vapour*.

A vapour is a gas that can be liquefied by applying pressure without cooling. Liquids like water, alcohol, ether and ammonia are vapours because temperatures of some two or three hundreds of degrees Celsius are needed to prevent liquefying. Gases like oxygen, nitrogen, hydrogen and methane have to be cooled a hundred or more degrees below Celsius zero before they can be liquefied. At ordinary temperatures they cannot be described as vapours.

Sublimation

This is the direct change of state from solid to vapour without going through a liquid stage.[*]

Solid carbon dioxide (carbon dioxide snow) looks rather like ordinary snow and sublimes at −78°C. Because it is so cold it can destroy (we say 'burn') the skin and it is for this reason that CO_2 snow or 'dry ice' is sometimes used to destroy warts.

LIQUIDS

Mixing a solid or liquid with a liquid

If a little salt, like common salt (sodium chloride), is mixed into water, it dissolves. The salt is the *solute* and what it dissolves in is called the *solvent*. The salt spreads evenly throughout the solvent and will not settle out. If the resulting solution is made to evaporate, it is the water that leaves and only the sodium chloride solid is left

[*]Usually the loss of molecules from the surface of a solid is very small even for those with a noticeable smell but some solids, frozen carbon dioxide being the best known example, evaporate readily so that, at a sufficiently high temperature (the sublimation temperature) normal atmospheric pressure is unable to prevent rapid sublimation.

behind, i.e. the well-known salt crystals. The salt has not been changed, by being dissolved, into anything else.

There is a limit to how much solid can be dissolved in a solvent, 35 g of sodium chloride in 100 g of water being the limit for a temperature of 20°C. A solution of maximum concentration is said to be a *saturated solution*. Increase of temperature enables a more concentrated solution to be obtained or a quicker rate of dissolving. In such a solution the solute molecules or ions mix freely with the solvent molecules, for in many cases the solute is ionised, sodium chloride, for example, being Na^+ and Cl^- ions.

Mixing liquid like alcohol into water also produces a solution. However, there is no limit in this case to the proportions of alcohol and water, and the term *solvent* is used for the liquid present in greatest quantity. In other cases there may be a limit to the solubility as found with dissolving solids.

Organic solids and liquids (see Chapter 18) are very often insoluble in water and suitable solvents are usually themselves organic. Alcohols and acetone are frequently found to be suitable solvents.

Mixing of insoluble solids and liquids with a liquid can produce uniform mixtures, namely suspensions and emulsions respectively, as pointed out below.

Suspensions

A *suspension* consists of particles of solid evenly mixed into a liquid. The particles are sufficiently big for them to be removed by filtering the suspension through a filter paper (Fig. 9.12). Also the particles may settle to the bottom of the liquid because of gravity.

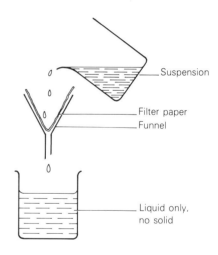

Fig. 9.12   Filtering a suspension

A paste

This consists of a powdered solid mixed with a small amount of liquid.

Colloids

These are often described as *colloidal solutions*. They behave very much like other solutions but are suspensions of particularly small particles of solid (comprising many molecules clumped together) or solutions with rather large solute molecules. Special filters can remove the particles from the solvent.

The very minute particles do not settle under gravity because they are continually being jostled by the thermal movements of the surrounding solvent molecules (an effect called *Brownian motion*). The particles are also electrically charged and the repulsion of the similar charges keeps the particles separated.

Examples of *colloids* are milk, a colloidal suspension of proteins in water, blood plasma and soap solutions. Some colloids can change from being fluid (when they are called *sols*) into jelly-like or semi-solid *gels* if they are cooled or chemically precipitated.

The gels probably owe their jelly-like consistency to the presence of fine threads of molecules which form a network through the gel. An example of this is gelatin (used for fruit jellies). Many cosmetics, such as foundation make-up, are available as gels.

Emulsions

Two liquids which do not mix, like oil and water, can under suitable conditions, produce an *emulsion* in which small droplets of one liquid are evenly spread throughout the other liquid. We can have an oil in water or a water in oil emulsion. Creams and lotions are emulsions.

To prevent the oil's separating out from the water and forming two separate layers of oil and water, chemicals called *emulsifying agents* are usually added to the mixture. These emulsifying agents are surface active agents and are discussed below. Soaps are examples of emulsifying agents.

Aerosols

Aerosols are very small solid particles or liquid droplets dispersed in air or other gas (a sort of gaseous colloid).

Foams

These consist of gas bubbles dispersed in a liquid.

Hydrates

(An explanation of hydrates is given in the footnote*, but some readers may prefer to go to the next section.)

Hygroscopic materials

Hygroscopic materials take water from the surrounding atmosphere. Hair and many fabrics have this property. Glycerin (glycerol) is also hygroscopic and is therefore included in many cosmetics as a means of keeping them moist. In other words, the glycerin acts as a *humectant*.

Distillation

Fairly pure solvent can often be obtained from a solution by making the solution evaporate and condensing the vapour in a separate container. If the solvent is much more volatile than the solute, then the condensed vapour is almost pure solvent. A simple distillation apparatus is shown in Fig. 9.13 (overleaf). Most of us know that

---

*Water molecules can loosely combine with some other molecules to form *hydrates*. This often happens when crystals are formed by evaporating a solution and the water combined in the crystals is called *water of crystallisation*. A good example is the combination of water with sodium carbonate (washing soda). Washing soda crystals tend to lose some of this water of crystallisation and become coated with the resulting anhydrous white powder. This loss of water of crystallisation is called *efflorescence*.

alcohol can be separated from water by distillation as illustrated by the distillation of spirits like whisky. Perfume oils, namely ethereal oils, are frequently obtained by *distillation* (see Chapter 21).

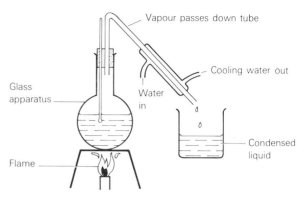

Fig. 9.13 A simple laboratory distillation apparatus

Distilled water

Tap water contains some dissolved chemicals, and these are sometimes a nuisance. Distillation of tap water produces water of much greater purity. Alternatively the offending chemicals can be removed with water softeners as explained in Chapter 18. *Distilled* or *deionised* water is particularly recommended for steamers.

Steam distillation

Instead of heating a liquid mixture directly to distil it, steam can be passed into it. Apart from other advantages, this provides a useful stirring up of the liquid.

Fractional distillation or rectification

This is the process in which several successive distillations of a liquid mixture are carried out in a column or tower. The most volatile component of the mixture passes through all the stages and emerges as a pure vapour at the top of the column while other components are separated into 'fractions' of different volatilities and can be taken from the various stages of the column. Some idea of the way this is done can be obtained by looking at Fig. 9.14.

Fractional distillation can be used for separating perfume oils. It is, however, better known for its use in the petroleum industry where crude petroleum oil is separated into various fractions including petrol, which is quite volatile, mineral oils used in cleansing creams (Chapter 21), and, least volatile, petroleum jelly (*petrolatum*) which is used frequently in many cosmetics as an emollient.

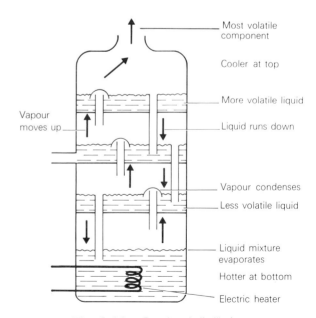

Fig. 9.14 Fractional distillation

Viscosity of fluids

Some liquids, and other fluids, flow more easily than others. For example, water flows more easily than treacle. A fluid that does not flow easily is said to be *viscous*. It has a high *viscosity*.

High viscosity is not simply due to a fluid's being heavy, having a high density (mercury is very dense yet flows very easily), but is due to molecules having difficulty in passing over each other. Large, long molecules, for example, are likely to cause high viscosity. The viscosity of a liquid is affected by its temperature, an increase of temperature often reducing the viscosity a lot.

Viscosity has to be considered carefully in the manufacture of many cosmetic preparations. Make-up which is too viscous may be too difficult to spread over the skin. An increase of temperature should not reduce the viscosity of a mascara or rouge so that it runs down the face when it is warm. Liquid shampoos should not run from the hand or from the head too easily. There must be time for it to be spread and worked into a lather. For similar reasons, cosmetics may be preferred as lotions rather than creams, a lotion having a lower viscosity enabling it to be spread easily. Lotions can also be dispensed easily from a variety of different containers. Emulsions made of oil and water typically contain a lower concentration of oil if a lotion is required rather than a cream. (See Figure 9.15.)

A lotion pours easily

A cream pours reluctantly

A thick cream or paste does not drip when applied by hand

Fig. 9.15 The viscosities of cosmetics

Surface tension

A drop of water hanging, for example, from a water tap (Fig. 9.16 (a)) behaves as if the water were contained within an invisible skin that holds it in. This effect is known as *surface tension*. The surface of the water is always trying to contract, i.e. is always in a state of tension. It is the surface that acts like a skin.*

(a)
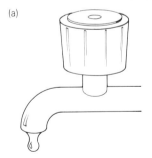

(b)
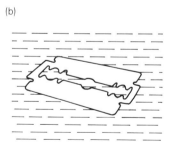

(c)

Fig. 9.16 Examples of the existence of surface tension: (a) drop of water; (b) razor blade floats on water; (c) film of liquid (e.g. milk) does not fall or sag at the end of a tube

*The cause of surface tension is that a water molecule at the surface has other water molecules close to it and attracting it on one side but far fewer molecules of air on its opposite side. Thus there is a net force inward, trying to take the molecule out of the surface. This explains why a surface is always trying to decrease its area. As a consequence, small drops of liquid falling through the air are formed into a spherical shape, the surface thus obtaining the smallest possible area.

Surface energy

(An explanation of this term is given in the footnote for those who are interested.*)

Interfacial energy

In an emulsion such as an oil in water emulsion there is a very large area of contact or *interface* between liquid and oil.

The potential energy of the interface (the *interfacial energy*) depends upon the surface energy of the oil and that of the water and is reduced by some attraction between oil and water molecules.

To help form an emulsion or to maintain (or 'stabilise') an emulsion formed by agitation of the mixture, it is necessary to reduce the interfacial energy. This is the function of an emulsifying agent. Cosmetic creams and lotions are emulsions and emulsifying agents are usually included in these.

Detergency (Chapter 22) depends upon mixing oily dirt or grease into water by forming an emulsion and, in fact, detergents are emulsifying agents.

Wetting a surface

(The molecular explanation is given in the footnote.†)

A liquid is said to *wet* the surface if it will spread over the surface rather than form droplets. Wetting can be helped by adding chemicals to reduce the surface tension of the liquid. Water wets glass but mercury does not (see Fig. 9.17).

(a)   (b)

Fig. 9.17 The wetting of a surface: (a) mercury on glass or water on a greasy surface; (b) water on clean glass

Surface-active agents

These are chemicals that, in small concentration, reduce surface and interfacial energies. They usually consist of fairly long molecules with one end, the head having a strong attraction for water while the tail (or tails) of the molecule has an affinity for fats and oils. If such a *surface-active agent* (or *surfactant*) is added to an oil in water emulsion, these molecules will become arranged as shown in Fig. 9.18, and act as an emulsifying agent.

Surface-active agents are used as emulsifying agents for many cosmetics; they are used as foamers and, as detergents, they are used for many cleaning purposes including some skin cleansers and most hair shampoos.

---

*In view of what has been said about surface tension it is clear that work has to be done, i.e. energy needs to be provided in order to increase the area of a liquid surface. This energy becomes potential energy of the surface and one can say that surface tension is caused by the 'desire' of the surface to reduce its potential energy.

When a foam is obtained, from a foam-bath liquid perhaps, there is a vast area of liquid–air surface. We should expect this to need a large supply of energy and so be difficult to obtain. In fact the surface tension of the water, and so the surface energy, has to be reduced to obtain a foam.

†When a liquid is placed in contact with a solid (or liquid) surface, the forces between the liquid molecules may be much greater than those between the liquid and solid so that the liquid remains separate from the surface as much as possible (Fig. 9.17 (a)). If the opposite is true, the attraction of the liquid to the surface encourages spreading of the liquid over the surface and a small angle of contact is obtained as shown in Fig. 9.17 (b).

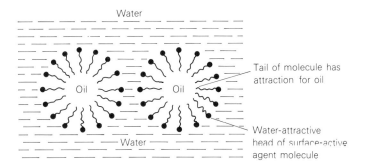

Fig. 9.18   The emulsification of oil droplets in water

Capillary rise (capillarity)

Fig. 9.19 shows a glass tube immersed in water. Where the water surface in the tube touches the glass there is strong attraction between the water and glass. In the remainder of the water surface the surface tension makes the surface try to contract so that it lifts the water up the tube. The *rise* will continue until the water reaches the top or until the weight of water in the tube equals the upward pull on the water.

This *capillary rise* is most noticeable in narrow tubes and it is this effect that enables blotting paper, various textiles that are wetted by water, and powders that have narrow gaps between the particles, to soak up water and other liquids.

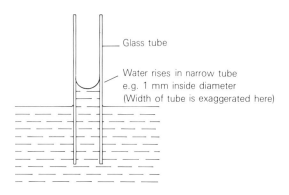

Fig. 9.19   Capillary rise in a glass tube

Diffusion

If two fluids that can mix are brought into contact but are not stirred or disturbed, the continual movements of molecules in one fluid cause some molecules to enter the other fluid. The process is called *diffusion* and is greater at higher temperatures. It causes gradual spreading of one fluid into the other.

In the lungs oxygen passes into the blood by diffusion.

Osmosis

When a solution in water, perhaps a glucose solution, is separated from some water by a membrane (or skin) that allows the small water molecules to pass through it but not the larger solute molecules, then *osmosis* takes place (Fig. 9.20, overleaf). This means that water will gradually pass through the membrane into the solution.

By way of some sort of explanation we can say that water molecules are in greater concentration on one side of the membrane and diffuse into the stronger solution but glucose molecules cannot diffuse from one to the other side.

Osmosis accounts for the movement of water in many situations in plants and in the human body. It enables water to enter the roots of plants from the soil and is important, for example, in the removal of water from the intestine and from kidney tubules.

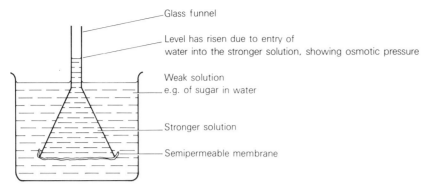

Fig. 9.20 Osmosis and osmotic pressure

—————————————————————— **Aerosol sprays** ——————————————————————

Three different kinds of aerosol spray device are shown in Fig. 9.21.

The first of these uses a flow of carbon dioxide gas to pick up and carry fine droplets of solution into the air.

The perfume spray relies upon the squeezing of a bulb to expel air which carries the perfume.

The aerosol can contains a very volatile liquid mixed with, for example, hair lacquer, and this propellant liquid keeps a high pressure of its vapour in the space above the liquid mixture. It is the vapour that pushes out the hair lacquer. (The lacquer is a solution of shellac in alcohol and the shellac dries as a film on the hair.)

(a)

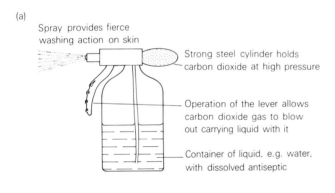

(b)

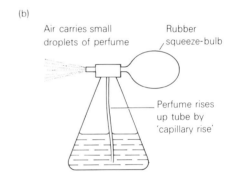

(c)

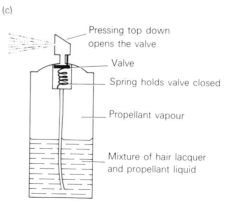

Fig. 9.21 Aerosol sprays: (a) using high-pressure carbon dioxide; (b) ordinary perfume spray; (c) an 'aerosol can' spray

The water cycle

It would be realistic to assume that a beauty therapist is interested in rain only to the extent of deciding whether an umbrella is needed; but it is rain that provides the water which arrives at the clinic and its journey decides what chemicals are dissolved in it.

Fig. 9.22 shows where rain comes from and where it goes to.

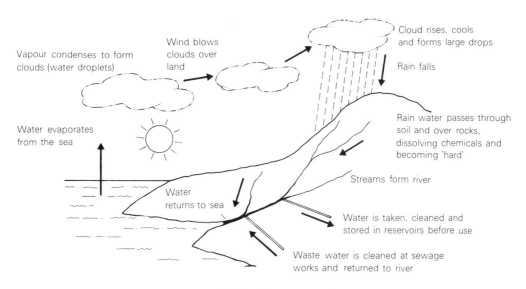

Fig. 9.22   The water cycle

MOISTURE IN THE AIR

Humidity

Humidity simply means the 'dampness of the air'. For comfortable working in a salon the humidity should be neither too high nor too low.

Air at 22°C can hold 20 g of water in every cubic metre. At this humidity the air is *saturated*. Any higher humidity would soon be reduced by condensation so that this is really the maximum possible. It is too high for comfort. A more acceptable level would be 10 g per m³ at this temperature.

It is usual to describe humidity by the value of the relative humidity (RH):

$$RH = \frac{\text{Number of g per m}^3 \text{ of water in the air}}{\text{Number of g per m}^3 \text{ of water that would saturate the air at the same temperature}}$$

The RH value is usually multiplied by 100 to give the RH as a percentage. As an example, 10 g per m³ at 22°C means an RH of $\frac{10}{20} \times 100\%$ which is 50%.

An RH of more than 65% makes one feel sticky and unpleasant when working and, if the RH is less than 30%, the air is too dry and can lead to dry and sore throats and coughs.

The RH in a Turkish (steam) bath might be 95% while for a sauna, where wooden walls absorb a lot of water vapour, the RH might be only 10%.

Dew-point

This is the temperature which can be quoted to describe the humidity of the air. It is the temperature to which the air must be cooled to make its water start condensing.

Air at 22°C with RH of 50% contains 10 g per m³ of water which is enough to saturate the air at 11°C. Thus 11°C is the *dew-point* of this air.

If warm air in a salon meets a surface at or below the dew-point, condensation will occur. This may happen on windows which are exposed to low temperatures outside. Air above a sink of hot water can become very humid and walls and a nearby mirror at room temperature may be cool enough to cause condensation and will soon be dripping wet (Figure 9.23).

Excessive condensation anywhere in the salon may give rise to rotting of woodwork, moulds on wall-paper and fabrics, discoloration of decor and smells, in addition to other inconveniences of dampness or puddles of water. The answer is to produce less water vapour from sinks, baths, showers, etc., or to remove vapour-laden air by very effective ventilation, or both. The rate of evaporation from baths, etc. may amount to some kilograms per hour. Even a sleeping person releases about 50 g per hour via the breath and perspiration.

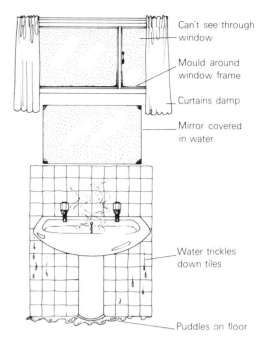

Fig. 9.23   Condensation from the air

Hygrometers

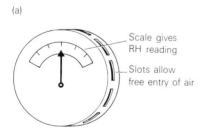

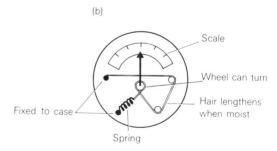

Fig. 9.24   The hair hygrometer: (a) view of outside;   (b) interior of hygrometer

For measuring the humidity of the air an instrument called a *hygrometer* is used.

One of the simplest designs of hygrometer is the hair hygrometer which makes use of the property of hair to lengthen slightly when it is moist and contract again when it dries. The working of this type of hygrometer can be seen in Fig. 9.24. The hair hygrometer is very suitable for permanent fixing on the salon wall although it is not very accurate.

The *wet-and-dry hygrometer* uses two thermometers, usually mercury-in-glass, which are fitted close together in a frame but with the bulb of one of the thermometers covered by a thin sleeve of wet muslin. Because of cooling by evaporation, the wet thermometer gives a lower reading than the dry thermometer which, of course, reads room temperature. This evaporation is greater if the air is dry and less when the humidity increases.

The relative humidity corresponding to the recorded wet-and-dry temperatures can be read from tables that are provided with the instrument.

Fig. 9.25 shows two wet-and-dry hygrometers. The whirling type gives more reliable results. Both need filling with water from time to time.

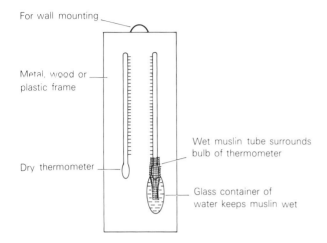

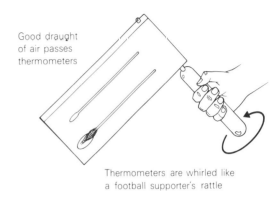

Fig. 9.25   Wet-and-dry-thermometer hygrometers

SUMMARY

- Change from solid to liquid is melting and the reverse is freezing or solidification. The change from liquid to vapour is evaporation and the reverse is condensation.

- Solid carbon dioxide (dry ice) sublimes, changing directly to gas without becoming liquid.

- Melting usually occurs at a definite temperature called the melting point and requires the supply of heat (which becomes the latent heat in the liquid).

- Crystalline substances melt at a very definite temperature. Amorphous and glassy substances (this includes many waxes) are not crystalline and melt over a range of temperature.

- Freezing usually causes a volume decrease.

- Freezing of water is very unusual — it causes an expansion. Freezing of water can burst pipes.

- Salt water remains liquid even at temperatures well below 0°C.

- Ordinary evaporation occurs only at the surface of a liquid. It occurs at any temperature but more quickly at higher temperatures.

- Boiling is escape of vapour, usually very rapid, from all parts of a liquid, as bubbles of vapour. It occurs at a definite temperature called the boiling point.

- Ordinary evaporation is encouraged by warmth and wind.

- Evaporation into a closed space will cause saturation and then no greater amount of vapour can be put into the space.

- The pressure of a saturated vapour is greater if the liquid is very volatile.

- Evaporation requires heat to be supplied (becoming latent heat of evaporation).

- Evaporation produces cooling.

- Cooling by evaporation can be used for freezing the skin to kill pain. Sweating cools the body by evaporation of the sweat. Many cosmetics produce cooling by evaporation of alcohol or water in the cosmetic.

- Steam is invisible water vapour or is visible steam consisting of small droplets of condensed vapour.

- Pressure discourages boiling, i.e. raises the boiling point. This fact is used in autoclaves.

- A vapour is a gas that can be liquefied without cooling it by applying pressure.

- Warming a liquid helps it to dissolve a solid.

- The dissolving solid is called the *solute* and the liquid in which it dissolves is the *solvent*.

- Evaporation of a solution leaves the solid behind.

- A suspension consists of such small particles of solid mixed into a liquid that the particles do not readily settle.

- A colloid differs from an ordinary suspension because very special filters are needed to separate solid from a colloid.

- Gels are jelly-like and are formed from some colloids.

- Emulsions comprise small droplets of a liquid in a second liquid in which it does not dissolve.

- Aerosols are minute liquid droplets in a gas such as air.

- Hygroscopic materials take water from the air, for example, glycerin (glycerol) is used as a humectant (to keep cosmetics moist).

- Distillation means heating a solution or a mixture of liquids so that the more volatile liquid evaporates first and is condensed. It separates the volatile liquid from the remainder.

- Distilled water is very pure.

- Fractional distillation enables many components of different volatilities to be separated from a mixture of liquids.

- Viscosity is the reluctance of a fluid to flow. Treacle is very viscous but water is not.

- Surface tension is the desire of a liquid surface to contract. It causes water to rise up a narrow tube and small drops of liquid to become round.

- Interfacial energy is concerned with the difficulty of obtaining a large area of surface between, for example, oil and water.

- Wetting of a surface means that a liquid spreads over the surface in its attempt to get as close to the surface as possible.

- Surface-active agents reduce surface tension and surface energies and assist wetting.

- Diffusion means one liquid or gas soaking into another.

- Osmosis is the movement of a liquid like water from a weak solution through a semipermeable membrane into a stronger solution.

- Humidity is the dampness of the air.

- Relative humidity is the mass of water in some air divided by the mass of water that this amount of air could hold. This fraction is multiplied by 100 to give %.

- Dew-point is the temperature to which some air has to be cooled to make it start condensing.

- To measure humidity a hygrometer is used (not a hydrometer).

- Water evaporates from the sea, forms clouds of small droplets which subsequently produce large droplets that fall as rain. The rain-water passes over soil and rocks and dissolves chemicals which make it hard. This water then forms streams and rivers that return the water to the sea. Thus the water travels around a cycle, the water cycle.

EXERCISE 9

1. What is meant by *latent heat?*

2. What is meant by *melting point?*

3. What is the effect on the melting point of water if salt (sodium chloride) is added?

4. Distinguish between *ordinary evaporation* and *boiling.*

5. Explain why water pipes sometimes burst in cold weather.

6. What should a beauty therapist do when a burst water pipe is discovered? (See Chapter 2.)

7. What is a *volatile liquid?*

8. Why do preparations containing alcohol feel cold when applied to the skin? (CGLI)

9. What is *steam?*

10. What is a *vapour?*

11. Distinguish between an *ordinary suspension* and a *colloid.*

12. Define the term *emulsion.* (CGLI)

13. Distinguish between an *emulsion* and an *aerosol.*

14. What is the purpose of using an emulsifying agent in a skin cream? (CGLI)

15. Explain the term *viscosity.*

16. Give an example of water showing that it has a surface tension.

17. What is a *surface-active agent?* What practical use can it have?

18. Distinguish between *diffusion* and *osmosis.*

19. Explain the meanings of the terms *humidity* and *relative humidity.*

20. Name two disadvantages of high humidity in the clinic, and one disadvantage of low humidity.

21. Define *dew-point.*

22. Name two problems arising from condensation of water in the clinic.

23. What is the purpose of: (a) a hydrometer, (b) a hygrometer? (CGLI)

24. At what stage in the water cycle does water acquire hardness?

25. The number of grams of water required per cubic metre to saturate air is shown in the following table.

Temperature/°C	−3	−2	−1	0	1
g per m³	3.9	4.2	4.5	4.8	5.2

Temperature/°C	2	3	4	5	6
g per m³	5.6	6.0	6.4	6.8	7.3

Temperature/°C	7	8	9	10	11
g per m³	7.9	8.3	8.8	9.5	10.1

Temperature/°C	12	13	14	15	16
g per m³	10.8	11.5	12.2	13.0	13.9

Temperature/°C	17	18	19	20	21
g per m³	14.9	16.2	17.0	18.0	18.8

Temperature/°C	22	23	24	25
g per m³	19.8	21.0	21.8	23.0

(a) If the temperature is 23°C, how much water can be held as vapour in 1 m³ of air?

(b) At this temperature how much vapour could be held in a room of volume 100 m³?

(c) If the relative humidity in a room is 50% and temperature is 25°C: (i) how much water vapour is contained in each cubic metre? (ii) what is the dew-point of this air?

(d) What is the relative humidity of air at 23°C if it contains 14 g of water per m³?

(e) If the dew-point of air in a room is −1°C, what is the relative humidity of the air at 20°C?

10

HEAT TRANSFER, HEATING AND VENTILATION

HEAT TRANSFER

Heat

Heat is energy that is moving from a hotter place to a colder place (because of the temperature difference) and there are three main ways in which this transfer can occur, namely:

- conduction
- convection
- radiation

Conduction of heat (thermal conduction)

This means heat is being *passed* from one part of a material to the next part, in fact, from one atom or molecule to the next.

Conduction can be nicely demonstrated with the apparatus shown in Fig. 10.1.

As seen in the diagram, heat is conducted quickly along the copper rod and the wax melts so that the wire coil falls off. Copper is a good conductor of heat.

Conduction occurs less easily in the other metal rods so that the heat spreads along the rods more slowly. Conduction of heat through glass is much slower.

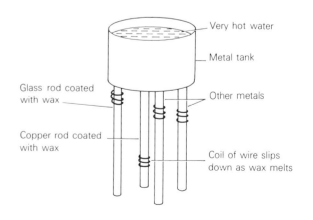

Very hot water

Metal tank

Other metals

Glass rod coated with wax

Copper rod coated with wax

Coil of wire slips down as wax melts

Fig. 10.1   A demonstration of thermal conduction

Good conductors of heat

Metals, both solid and liquid, are good *conductors* of heat, some being better than others as shown in the experiment above.* The ease with which heat spreads through mercury enables the mercury-in-glass thermometer to give a reading quickly.

Heat insulators (bad conductors)

Glass, plastics, cloth, still air or water and brick are poor conductors of heat and are called heat *insulators* because they can reduce escape of heat from hot objects and stop heat reaching cold objects.

Some consequences of heat conduction

A person walking barefoot on a floor which is a good conductor will get cold feet because heat travels easily from the foot into the floor. Similarly the couch on which the beauty therapist's client is to lie could feel cold unless a blanket and sheet or similar heat insulation is placed over it. A bare metal electrode applied to the skin could feel cold for the same reason.

The walls of a steam bath can be very hot and, if the bath is made of metal and comes into contact with a client's body, the heat will quickly leave the metal and burn the skin. The metal bath can be covered with a towel to avoid this. The towel that is placed around the neck also stops the escape of steam, as in Fig. 9.11 (p. 90).

Wax-heating pots, like cooking pots, where heat has to enter from outside, must be made of a good thermal conductor and so they are made of metal. However, the handle of the pot will feel too hot to touch if it also is made of metal, so an insulating handle is fitted. Otherwise it can be handled using gloves or a cloth.

The escape of heat from water pipes into cold surroundings is reduced by lagging the pipes with thermally insulating material such as fibre-glass wool or a mineral wool (both of which are poor conductors themselves and also enclose a lot of still air). The success of any heat insulation depends partly on how well the chosen material conducts (described by its 'thermal conductivity') and also on the thickness of the material that is used.

The use of thermal insulation in the construction of a sauna is seen in Fig. 10.2.

Yet another example of thermal insulation is the

Wood panel, inner surface of wall

Thick layer of fibre-glass wool or rock wool

Supporting frame

Wood panel, outer surface of wall

Fig. 10.2   The thermal insulation in the wall of a sauna

*One way in which heat is conducted in a solid is as follows. At a hot place atoms are vibrating a lot and they shake neighbouring atoms so that these atoms are made to move more and thus the temperature rises in the neighbouring area. So energy has spread from the hotter to the cooler place next to it. In metals, conduction by this process is helped by a further process due to movement of free electrons. The extra conduction by free electrons ensures that metals are good conductors.

combination of glass and still air to reduce loss of heat from a room when double glazing is installed. If the two layers of glass have a vacuum between them instead of air, then conduction through the window will be even less.

Clothing the body is really another example of lagging, and clothes which trap large quantities of air are very effective. Wool fibres have a scaly surface that traps air so that wool fabrics will keep the body warm. Silk, like wool, is a protein material and is a poor conductor.

Linen, cotton and viscose rayon are comparatively good conductors. Subcutaneous fat is a poor conductor and reduces unnecessary heat loss from the body.

During wax treatments the time taken for the wax to cool must be decided partly by the thermal conductivity of the wax itself. Sometimes after wax is applied, for example, as a heat treatment of a limb, it is covered with a sheet of foil over which a thickness of towel is placed to reduce escape of heat.

Convection

In *convection* heat is *carried* by the movement of a liquid or gas, air for example. The convection may be forced convection or natural convection depending on how the movement of the fluid is caused.

Natural convection

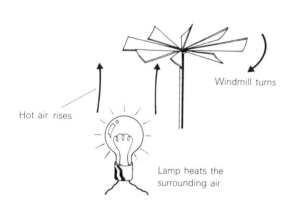

Hot air rises

Windmill turns

Lamp heats the surrounding air

Fig. 10.3 A demonstration of natural convection — the convection windmill

When the convection is *natural convection* the flow is caused by the heat that is being transferred. Fig. 10.3 shows how natural convection can be demonstrated.

The lamp, or other hot object, heats the air close to it making it become less dense. The hot air therefore rises. Cooler air takes its place and, in time, this is heated and rises. Thus natural convection causes a circulation of air called a convection current. It is natural convection that is responsible for the statement 'heat rises'.

The convection windmill is sometimes used with a red lamp in a heater to create the effect of flickering flame.

Convection is a very effective method of heat transfer because the air movement causes the hot object to be exposed to cold air at all times, carrying heat away as fast as it leaves the hot object.

Forced convection

If the air or other fluid that carries heat from the hot object is made to move, for example, by a fan or pump, then the convection is described as *forced* (see Fig. 10.4).

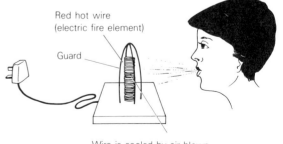

Red hot wire (electric fire element)

Guard

Wire is cooled by air blown over it and its redness is reduced

Fig. 10.4 A demonstration of forced convection

Making use of natural convection

An ordinary convector heater allows air to enter the bottom of the case and it is heated by wires (usually almost red-hot) that are made hot by electric current flowing through them. The hot air rises and leaves through an opening at the top of the case (Fig. 10.5). Note that the air inlet and outlet areas must not be obstructed at all for fear of the appliance becoming overheated inside.

tion. Radiation is explained later in this chapter. The heat is supplied to the radiator by hot water or steam passing through it or by electric heating of oil inside the radiator.

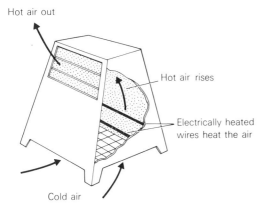

Fig. 10.5   A room heater of the ordinary (natural) convector type

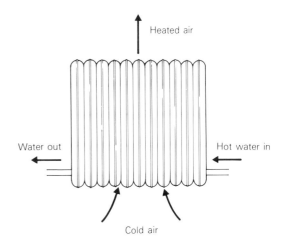

Fig. 10.6   A radiator depends a lot upon natural convection

In spite of its name, a radiator gives heat to a room approximately half by convection and only half by radia-

Bearing in mind that 'heat rises', we expect to find the hottest air close to the ceiling of a room. This is true also of a sauna such as the small, indoor sauna cabin illustrated in Fig. 10.7.

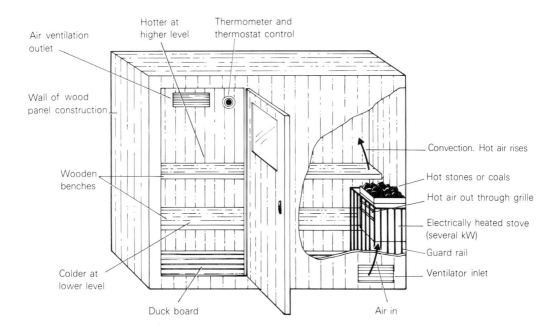

Fig. 10.7   A typical indoor sauna

In the sauna shown the stove is electrically heated. In other saunas, gas or even log-burning stoves are used. Heat from the stove rises and reaches other parts of the cabin by natural convection. When someone is lying on a higher bench, the temperature experienced is higher than on a lower bench. The walls are insulating because of the wood panel and rock wool construction, or because of the thickness of wood used in the case of log saunas. The wood also reduces the humidity so that the sauna provides a 'dry heat', but the humidity can be quickly increased by sprinkling some water over the hot stones on the top of the stove.

Natural convection even plays its part in the heating of water in a kettle. Heat is supplied at the bottom but is distributed to all parts of the water by convection. The circulation of water by natural convection is particularly useful in domestic hot-water systems. A diagram of a typical system is seen in Fig. 10.8.

In this system water needs to be heated and then stored ready for use while other water is heated. The flow required for this is achieved by natural convection.

Water that has been heated in the 'boiler' (it should not normally boil) rises because of its lower density and enters the top of the storage tank while colder, denser water moves down into the boiler. In the hot water storage tank the hot and cold water remain separate because convection will only move hot water upwards and water conducts only poorly. Water will also circulate by convection through the radiator. When a hot tap is used in the bathroom or kitchen, water comes out because the tap is below the water level in the cold tank, and the water is hot because it comes to the tap from the top of the hot tank.

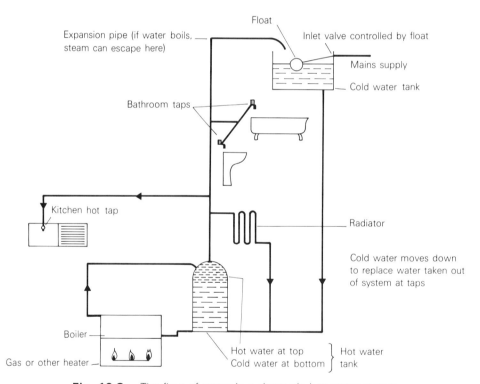

Fig. 10.8   The flow of water in a domestic hot water system

Some uses for forced convection

Although forced convection can be a nuisance when a draught of air removes heat from our bodies and we complain of a cold draught, it can be usefully employed, not least in the beauty salon.

The diagram in Fig. 10.9 (a) shows how a room heater, called a *fan heater*, blows air over hot wires and then into the room. Quite large amounts of heat can be handled (per second) and the hot air can be directed to the

108

required part of the room. A hair-dryer works on the same principle (Fig. 10.9 (b)).

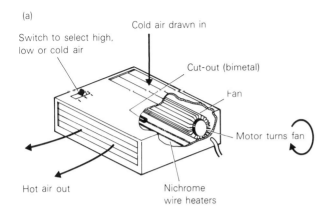

(a)

Switch to select high, low or cold air

Cold air drawn in

Cut-out (bimetal)

Fan

Motor turns fan

Hot air out

Nichrome wire heaters

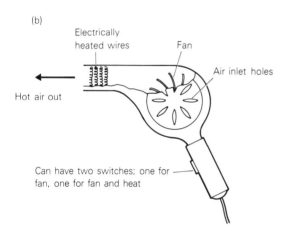

(b)

Electrically heated wires

Fan

Air inlet holes

Hot air out

Can have two switches; one for fan, one for fan and heat

Fig. 10.9 (a) A fan heater; (b) a hair-dryer

Storage heaters (see Chapter 7) may have a fan built in to obtain fast removal of heat by blowing air between the storage elements and sending the hot air out into the room through a grille or slots in the case.

For central heating, where the heating of several rooms is obtained with one heater, the heat may be fed to each room as hot air from an electric, gas or other heater. Wide pipes or ducts are needed to carry the air which is driven by a fan. Alternatively radiators can be used, fed with hot water usually pumped by an electrically-operated pump, through a 'small-bore system', so-called because of the small diameter piping used.

A pumped central heating system and the body's blood system are not unalike. Each uses forced convection,

pumping a liquid to provide heat where it is needed. The pumping of the blood is largely done by the heart. Blood also removes heat from any part of the body where heat is being produced rapidly, such as in muscles during exercise.

Fig. 10.10 shows how convection is used in gas heaters of the balanced flue type. With almost all gas-burning heaters the chemical products from the burning of the gas must be released outside of the building so that a flue is needed but, in the balanced flue design, the air needed for the gas to burn is also taken in through the flue. In this way the room air is kept separated from the gas burning.

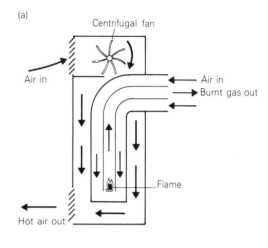

(a)

Centrifugal fan

Air in

Air in
Burnt gas out

Flame

Hot air out

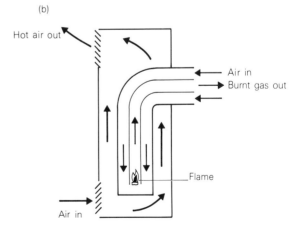

(b)

Hot air out

Air in
Burnt gas out

Flame

Air in

Fig. 10.10 Balanced flue gas burning heaters:
  (a) fan-driven; (b) natural convection

Radiation

It is pleasant to think of sitting in front of a blazing fire or watching the glow from red-hot coals and feeling the warmth on one's face or out-stretched hands. The heat is received as radiation.

Radiated heat is heat that has been thrown out (or 'emitted'), not conducted through the air, nor convected by the air since convection is carrying heat upwards. In fact, if there were no air between the fire and the person, there would be just as much radiated heat.

The nature of this invisible, radiated heat is discussed in Chapter 17 where this radiation is called *infra-red radiation* (abbreviated as IR).

Radiated heat is only received in sufficient quantity to be easily noticeable if the heater, or other 'source' of radiated heat, is very hot or is of large area.

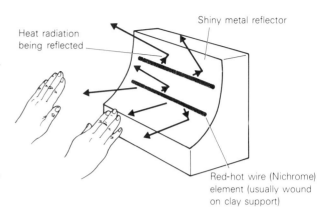

Fig. 10.11 Radiated heat

Reflection of radiated heat

We are all familiar with the reflection of light from mirrors and other shiny surfaces. *Radiated heat* is reflected by shiny surfaces too. Fig. 10.11 shows how a reflector is used as part of a radiant electric fire. Heat that is emitted from the hot elements away from the required direction is reflected forward.

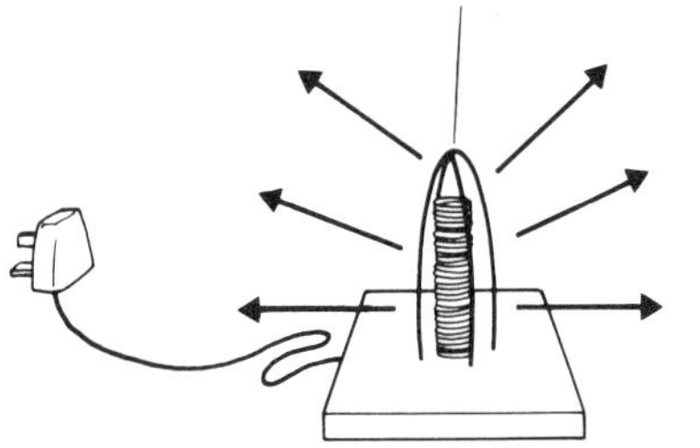

White surfaces are quite good reflectors too, but dark surfaces and matt surfaces, especially a matt, black colour, are good absorbers of radiated heat. A good absorber soaks up the radiated heat and it becomes ordinary heat (or rather internal energy) of atom vibration. Thus dark surfaces reflect little heat.

The experiment illustrated in Fig. 10.12 confirms that a black surface is a good absorber and poor reflector of heat because the blackened thermometer shows a higher temperature:

Fig. 10.12 An experiment to compare the absorption of radiated heat by different surfaces

To summarise:

Black absorbs. White reflects.

Radiation by different hot surfaces

Fig. 10.13 shows apparatus that can be used to study the radiation of heat from various surfaces.

The metal can has faces of identical size but different colours and textures. Each face in turn is shown to the detector. The highest meter reading is obtained with the matt, black surface, the lowest readings from the shiny, mirror surface and gloss white, painted surface.

We now see that:

Black is a good absorber and, when hot, a good radiator. White is a good reflector and, when hot, a poor radiator.

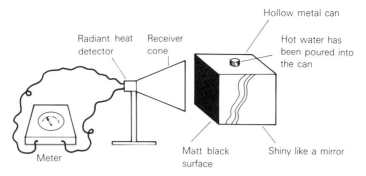

Fig. 10.13 An experiment to compare the radiation of heat from various surfaces

Some examples of radiation

We have already said that an ordinary room-heating radiator gives out perhaps half of its heat by radiation. Ideally these radiators should be matt black in colour but the use of other colours, even light colours, is often preferred. In contrast to this, neither a kettle nor a wax heater is meant to warm the surrounding room air and radiation may be reduced to a minimum by using polished metal exteriors, perhaps shiny and chromium-plated so that tarnishing is minimised as well.

Vacuum flasks can be useful for a variety of purposes, either keeping the contents hot or keeping them cold. As shown in Fig. 10.14, heat is prevented from leaving the enclosed hot liquid by conduction or convection because of the vacuum.

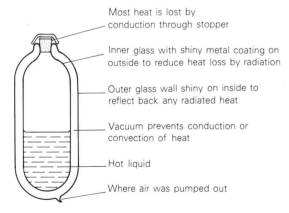

Most heat is lost by conduction through stopper

Inner glass with shiny metal coating on outside to reduce heat loss by radiation

Outer glass wall shiny on inside to reflect back any radiated heat

Vacuum prevents conduction or convection of heat

Hot liquid

Where air was pumped out

Fig. 10.14 A vacuum flask

As regards radiation there is a polished, 'silvered' outer coating (probably aluminium) which may become warm but is a poor radiator. The other silvered surface is

a good reflector that returns the larger part of any radiated heat that crosses the vacuum. Most of the heat lost will escape through the stopper.

For clothing, few people will wear shiny metallic film to reflect the radiated heat from the sun but white clothing does offer coolness in the summer (Fig. 10.15).

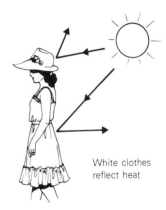

White clothes reflect heat

Fig. 10.15 Keeping cool in the summer

Even our safety may depend on understanding radiation of heat. A piece of metal with a dark surface, like the drain cover in Fig. 10.16 (overleaf) may become very hot in the sun because it absorbs heat radiation so well. The metal is a good conductor and so its high temperature is easily and painfully detected by bare feet.

This illustration may seem frivolous but a metal object, especially if dark in colour, left beneath an infra-red lamp of the kind used for therapy purposes would get very hot and, if touched, this good conductor would *feel* very hot.

Fig. 10.16 Dark surfaces can become very hot

The gas fire

Use is made of radiation in *gas fires*. The gas flames heat the projections (like pimples) on the clay radiant elements. These parts of the elements glow red hot and radiate heat directly into the room.

As seen in Fig. 10.17, advantage is often taken of convection by letting air flow behind the radiating elements and past the flue. Heat passes into this air which consequently rises and flows out through a grille.

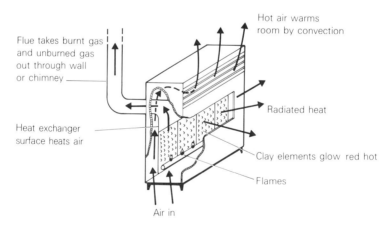

Fig. 10.17 A gas fire

Heat transfer by evaporation

To complete the discussion, we must mention a fourth way in which heat is transferred — by *evaporation*. Evaporation produces cooling, and several examples of this were described in Chapter 9. Heat loss by this method can be very effective. It can be regarded as con-vection because the heat is carried away as energy of the vapour, in fact as latent heat. Condensation of vapour delivers heat to the cool surface where the vapour condenses.

Comfort in the salon

If clients and staff in a salon are to feel comfortable the air must satisfy the following requirements.

The *temperature* must not be too high or too low, 22°C perhaps being ideal. This means that heating must be provided, maybe local heating by means of electric convectors or central heating such as hot-water-fed radiators. Central heating for background heat with additional local heating that can quickly be brought into use when needed is a nice idea. Heating that is controlled thermostatically is most convenient and may prove more economical.

Heat is lost by conduction through walls, ceilings and particularly single-glazed windows as well as by replacement of warm air by cold when doors or windows are opened. It may therefore be necessary to provide heating at a rate of several kilowatts even for rooms that are not large. It might be noted here that electrical equipment produces heat when in use and also people produce about 100 watts each of heat.

A typical heating system is shown in Fig. 10.18.

Humidity should neither be so high as to make the occupants of the room sweat nor so low as to cause dryness of the mouth, nose or throat. A relative humidity of 30–65% is suitable.

Air movement is necessary for comfort, an air velocity of about 200 mm per second being desirable. Too little air movement makes a room seem stuffy and too great an air velocity makes the room feel draughty. A number of different instruments are available for measuring air speeds in rooms.

Cigarette smoke, chemical fumes and unpleasant smells must be avoided or their build-up prevented by good ventilation.

Adequate oxygen is clearly essential for breathing but no problem is likely to arise in this respect in a beauty salon even though expired air contains only about 16% of oxygen compared with 21% in fresh air.

Build-up of carbon dioxide due to respiration of people even in a crowded room is again not sufficient to become a problem, although expired air does contain about 4% of this gas compared with only 0.04% in fresh air.

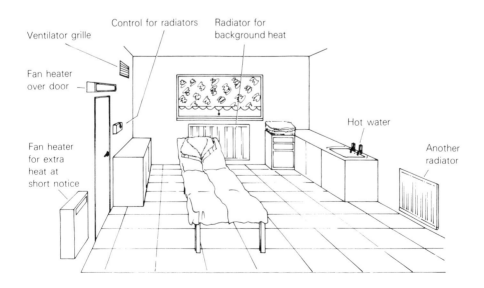

Fig. 10.18   Heating the salon

Ventilation

All of the above-mentioned comfort considerations are influenced by the room *ventilation*. In most salon areas, including those devoted to massage, electrotherapy and make-up and solarium and gymnasium rooms, the ventilation should, as a rough guide, provide 17 m³ per hour of fresh air per person. The same is true for a living room of a house. For a bathroom or similar area and for kitchens, more ventilation is required.

Natural ventilation

When no fans or pumps are used to cause ventilation we are left with *natural ventilation*. This may result from open windows or doors but it also occurs by movement of air through cracks and other gaps around doors and windows.

Air changes per hour

An *air change* for a room means the entry of fresh air equal in volume to the volume of the room.

In buildings of modern design one expects at least one air change per hour to occur even when doors and windows are closed. In older buildings where doors and windows mostly do not fit well, this natural ventilation will be more like two air changes per hour.

Fires which make hot air go up through a flue add to the natural ventilation.

Calculating how much ventilation is needed

Suppose that a particular room measures 8.5 m × 4 m × 2.5 m and is to hold 10 people requiring 17 m³ per hour of fresh air per person. How many air changes per hour are required?

The volume of this room is 8.5 × 4 × 2.5 m³ which equals 85 m³.

The fresh air required per hour is 17 × 10 which is 170 m³.

The number of room changes per hour is 170/85 which is exactly 2 per hour.

Methods of obtaining adequate ventilation

(1) *Opening some windows.* The disadvantages involved are possible cold draughts, inconvenience of curtains or papers blowing about, and letting in noise or smells from outside.

(2) *Ventilation using natural convection.* One example of air flow obtained in this way is shown in Fig. 10.19. The use of the hopper inlet makes the cold air move first into the hot air at the top of the room so that it warms a little before it falls down. Hot air rises and leaves through the ventilator grille.

This arrangement avoids cold draughts.

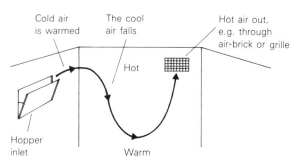

Fig. 10.19 Ventilation by natural convection

Fig. 10.20 (a) shows a Cooper disc which can be useful as an air inlet or outlet that can easily be opened or closed. Fig. 10.20 (b) shows a sash window where cold air can enter the lower opening and hot air leave at the top. Draughts are likely.

(a)

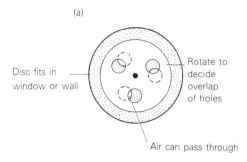

Disc fits in window or wall

Rotate to decide overlap of holes

Air can pass through

(b)

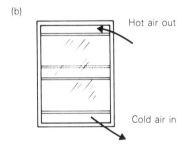

Hot air out

Cold air in

Fig. 10.20 (a) A Cooper disc; (b) air movement through a sash window

(3) *Artificial ventilation.* Here fans are used to produce the air movement. One immediate advantage is that the ventilation is not affected by how windy it is outside. The methods employed can be classified as:

(a) *Exhaust systems,* in which the fan pushes air out of the building.

(b) *Supply systems,* in which air is blown into the building by the fan and the air can be filtered to remove dust and 'air-conditioned' if required before it is allowed to spread through the building. This method can therefore produce air of suitable temperature independent of the temperature of the outside air temperature.

(c) *Balanced systems* that have one fan blowing air in with the advantages of the supply system and also another fan to send air out.

Several rooms or several air inlets or outlets can be served by one fan if the air is driven through ducts as suggested in Fig. 10.21.

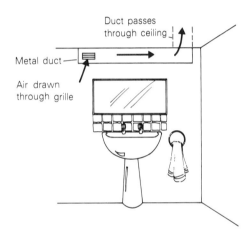

Duct passes through ceiling

Metal duct

Air drawn through grille

Fig. 10.21 A ventilation duct system or trunk system

Air-conditioning

Air-conditioning means filtering out dust, warming or cooling the air, and moistening or drying the air to give it an acceptable humidity.

An air-conditioning unit can be fitted to the inlet fan in a supply or balanced system. Alternatively, small independent air-conditioning units are available that can be fitted to the wall of any salon, as seen in Fig. 10.22.

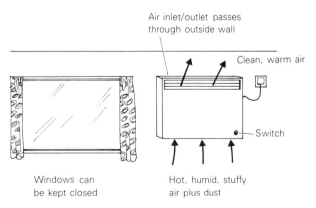

Air inlet/outlet passes through outside wall

Clean, warm air

Switch

Windows can be kept closed

Hot, humid, stuffy air plus dust

Fig. 10.22 A wall-mounted air-conditioning unit

Ventilation versus condensation

The problems that accompany heavy condensation of water from the air in a room have been stressed in a previous chapter. Unless evaporation from baths and showers can be reduced the answer is likely to be improved ventilation. Additional room heating or thermal insulation also help. Condensation on walls is most likely to be found where the air flow is least, such as in corners of a room. Cupboards are possible trouble spots. Air flow in these areas should not be obstructed.

SUMMARY

- Heat may be transferred by conduction, convection and radiation.

- Conduction means that heat is *passed* on (from molecule to molecule or atom to atom).

- Convection means that heat is carried by the movement of gas or liquid.

- In natural convection the flow of liquid or gas is caused by the heating.

- In forced convection the gas or liquid is made to flow by a fan or pump.

- Radiation means heat being thrown, or emitted, from the surface of a hot object.

- Metals are good conductors of heat.

- Glass, brick, most plastics, still air, or even still water, are poor conductors of heat or thermal insulators.

- Poor conductors of heat are useful for insulation to prevent heat leaving saunas, water pipes, roofs, people's bodies.

- Good conductors of heat, in fact metals, are used for saucepans, pots for wax-heating and any pipes where heat needs to enter or leave easily.

- Good conductors of heat are cold to the touch.

- Room radiators, convector heaters and domestic hot water systems make use of natural convection.

- Fan heaters and hair-dryers use forced convection.

- A room or sauna is hottest near the ceiling because 'hot air rises'.

- Radiated heat can be reflected by mirror surfaces.

- White surfaces and shiny surfaces are good reflectors of radiated heat and, as hot surfaces, are poor radiators. Black surfaces are good absorbers and, as hot surfaces, are the best radiators.

- White clothes keep us cool in summer.

- For comfort the relative humidity of the air in the clinic should be between 30 and 65%.

- Fresh air contains 21% oxygen, 0.04% carbon dioxide. Expired air contains 16% and 4%. 17 m^3 of fresh air should be allowed per person per hour.

- Adequate ventilation may be obtained by open windows (draught problems) or using Cooper discs, louvres or hopper inlets. Instead of natural ventilation, fans can be used and it may then be convenient to include air conditioning.

EXERCISE 10

1. Name three processes by which heat may be transferred.

2. Suggest a suitable material for lagging pipes.

3. Why are metals cold to the touch?

4. Explain how heat travels from: (a) an electric fire, (b) a hot water radiator, (c) a fan heater.

5. Explain why the surface of a kettle should preferably be shiny.

6. Describe *two* problems associated with using natural methods to ventilate a salon. (CGLI)

7. Compare the percentage volumes of oxygen and carbon dioxide in inspired and expired air. (CGLI)

8. Which method of heat transference makes possible the natural ventilation of a room? (CGLI)

11

SOUND, NOISE AND VIBRATION

SOUND AND NOISE

What is sound?

'What is happening in the air through which sound travels from a person speaking to a listener's ear?'

A more practical question might be 'What is happening in the soft body tissues into which vibrations are sent by a vibrating massage apparatus?' The latter question may seem more relevant to the reader, but it will be easier and useful to consider *sound* first.

The source of a sound, whether it be a person's voice, the cone of a loudspeaker or the earpiece of a telephone, etc. contains a surface which moves backwards and forwards repeatedly, i.e. a vibrating surface. As this movement occurs the air adjacent to the surface is alternately compressed (squeezed) and rarefied. One might almost say 'squeezed and stretched'. The air pressure is a little greater than atmospheric pressure where the air is compressed and a little less than atmospheric where there is a rarefaction.

Sound consists of compressions and rarefactions in the air travelling out from the speaker or other source. This is illustrated in Fig. 11.1.

Let's try to see why compressions and rarefactions move away from the source. As soon as air is squeezed it begins to push out because it is not confined by a container at all. Consequently, compressions formed at the source move out from it. Similarly as soon as a rarefaction is formed at the source, nearby air will expand into this reduced pressure region so that there is a pressure reduction in the nearby area. The rarefaction has moved outward. A more complete explanation than this is beyond the scope of this book.

The whole system of compressions and rarefactions, which is the sound, moves forward so that the ear receives a C then a R and so on alternately, while the source is creating C's and R's.

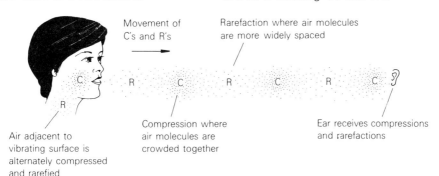

Movement of C's and R's

Rarefaction where air molecules are more widely spaced

Air adjacent to vibrating surface is alternately compressed and rarefied

Compression where air molecules are crowded together

Ear receives compressions and rarefactions

Fig. 11.1　Sound waves in the air

117

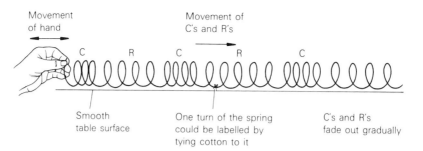

Fig. 11.2 Using a long spring to explain how sound travels

A good aid to understanding how sound travels is a long 'slinky spring'. If one end is made to move to and fro as shown in Fig. 11.2 then 'compressions' (where the turns of the spring are close together) and 'rarefactions' (where the turns are more widely spaced) move along the spring. This is just like C's and R's in the air.

If time could be stopped, like taking a photograph of C's and R's in a spring or drawing a diagram like Fig. 11.1 or Fig. 11.2, then the variation of air pressure with distance would look like Fig. 11.3.

This graph of pressure against distance explains immediately why a sound is often described as a *wave*. The pressure rises and falls as we look at increasing distances from the source and this is just like a photograph of waves on the sea.

Before leaving this discussion, consider what is happening at any one place where sound is present — for example, the place where we have the labelled part of the spring in Fig. 11.2. It is important to note that the labelled part simply moves to left and right alternately so that it finds itself alternately in a compression and a rarefaction. This emphasises that sound is a continuous forward movement of C's and R's but it involves only small vibrations of air. Air itself does *not* travel from the source to the receiver.

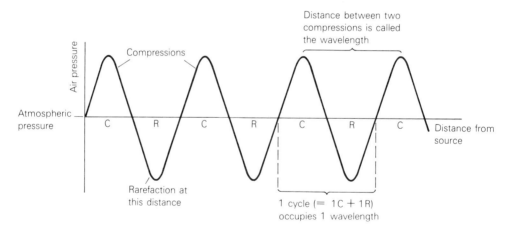

Fig. 11.3 Sound involves pressure fluctuations

Frequency, wavelength and velocity of sound

The *frequency* of a sound is simply the number of C's or R's produced per second (or cycles per second) and, as with electric currents, the *frequency* is written as so many hertz (Hz).

The *wavelength* of a sound is the distance between one C and the next or the distance occupied by one C and one R (i.e. by 1 cycle of the wave).

118

Sound travels at approximately 340 m per s in air. This is its speed (or velocity).

An important equation for sound is this:

Frequency × Wavelength = Velocity

Using the symbols f, λ (lambda) and v, this becomes

$$f \times \lambda = v$$

This equation is true for sound waves and for *all* types of wave (see Chapter 17).

It is easy to understand this equation as follows. Any place receiving sound waves receives f cycles, each of length λ, in each second. Thus, the $\lambda \times f$ is the number of metres of sound wave received per second.

The pitch of a sound

How high a note seems to the ear is called the *pitch*. A deep voice is a low pitch, a soprano can sing notes of high pitch and a whistle is of a higher pitch still. The pitch is decided almost entirely by the frequency of the sound:

High-frequency sounds are of high pitch.

The ear

Fig. 11.4 shows the main parts of the *ear*.

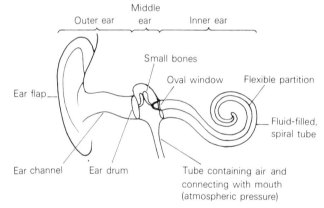

Fig. 11.4   The structure of the ear

When sound reaches the ear the air in the ear channel is alternately compressed and rarefied so that the ear drum is made to move in and out. This movement is linked by the small ear bones to the oval window. As this window moves it creates sound again, but this time travelling in fluid, which makes the flexible partition vibrate and this movement affects nerve fibres which send news of the sound arriving (in the form of nerve impulses) to the brain.

Audiosonic, ultrasonic and infrasonic waves

The ear (and brain) respond to sound waves with frequencies from about 50 Hz to about 15 000 Hz (or a little more for younger people). These frequencies are called *audiofrequencies* because they are audible and such sound can be called *audiosound* or *audiosonic*. *Ultrasonic* waves have frequencies higher than 20 000 Hz and *infrasonic* waves have frequencies below the audible range.

Sound caused by striking

Hitting an object will make it shake. The frequency at which it vibrates and the sound waves which it causes in the surrounding air, are decided by the size, shape and material of which the object is made and the sound soon weakens and fades away. *Striking* an object at a rate of say three strikes per second will produce sound but its frequency will not be 3 Hz.

Noise

Unwanted sound is called *noise*.

When sounds of several frequencies are heard at the same time the combined effect can be pleasant if the frequencies are suitably related (as when chords are played on a musical instrument). Unsuitable combinations are discordant and almost certainly are unwanted.

Sounds can also be unwanted if they are too loud, 'too loud to hear oneself speak' perhaps.

Loudness

If we consider a sound of a particular frequency, say 1000 Hz, its strength (or *intensity*) is dependent upon the sizes of the pressure changes of the sound. The *loudness* experienced by a person hearing it is decided by this intensity.

The loudness does not, fortunately for us, double when the sound pressure fluctuations are doubled in size, so that special methods of describing loudness are used, particularly the decibel method described below.

Another peculiarity of the ear is that it is less sensitive to low-frequency sounds.

Sound energy

Sound involves to and fro *movements* of air molecules. The air is squeezed in some places and rarefied in others. So where there is sound there is kinetic energy and potential energy.

Decibel (dB)

The quietest sound that can be heard is described as being zero *decibels* (dB) and a sound loud enough to be painful on the ears is about 120 dB. Where ordinary sounds fall in this range is indicated in Fig. 11.5.

For speech at 2 metres distance to be heard easily, other sounds being received should be much quieter, something like 10 dB lower, i.e. about 50 dB.

Conversely, to prevent conversation being heard at a distance of say 4 m some music could be played at about 50 dB or more.

Bearing in mind that low-frequency sounds and high frequencies are not heard so well by the ear as frequencies around 1000 Hz, this can be allowed for and the answer is recorded in dB with a letter A added. For example, 90 dB of sound at 100 Hz is no louder than 70 dB at 1000 Hz and so we can describe it as 70 dBA.

A dBA value of 40 or below is desirable for a clinic before clients arrive. A sound level meter can be used to measure level of sound (Fig. 11.6).

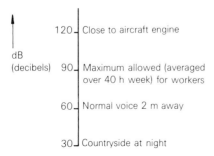

Fig. 11.5   Sound levels in decibels

120

dB
120 ─ Close to aircraft engine
(decibels) 90 ─ Maximum allowed (averaged over 40 h week) for workers
60 ─ Normal voice 2 m away
30 ─ Countryside at night

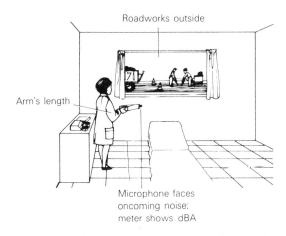

Fig. **11.6**   Using a sound level meter in the salon

Noise reduction in a salon

Sound waves cannot pass easily through heavy objects like brick walls and are reflected instead. In the case of objects small compared with the sound's wavelength, sound waves pass round the object. Sound will pass easily through thin curtains and the like, the light curtain vibrating just like the air.

The reflection of sound can be a nuisance because it can send noise round corners. But, a heavy wall can usefully reflect noise away and the reflection of sound backwards and forwards between the panes of glass of a double-glazed window enables repeated absorption of the sound to occur.

Sound can be reduced in strength by absorption (which means taking energy from the sound) if it is made to pass through a considerable thickness of porous material that contains fine channels in which movements of air are resisted by the viscosity of the air. Acoustic tiles or a few inches thickness of rock wool are effective absorbers of sound. Fig. 11.7 shows how noise can be reduced in the salon.

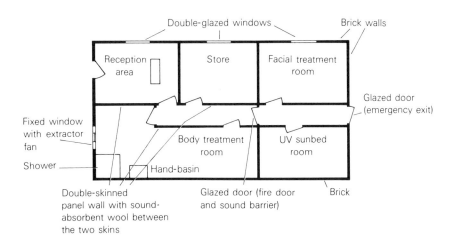

Fig. **11.7**   Noise reduction in the salon

Vibration massage

To-and-fro movements (*vibrations*) applied to the skin surface will cause compressions and rarefactions in the underlying soft tissues.

These movements encourage movement of the body fluids, lymph and blood, and cause beneficial erythema (reddening due to increased blood supply). Massage by vibration, in contrast with some other electrical apparatus,

produces no muscle action and does not itself cause any chemical effects.

Two kinds of *vibration massagers* are described below. There are vibrators which operate at the much higher ultrasonic frequencies which are used by physiotherapists but not apparently by beauty therapists.

Audiosonic vibrators

The appearance of a typical *audiosonic vibrator* is sketched in Fig. 11.8.

The applicator in contact with the body moves to and fro a few millimetres and the frequency is typically a hundred or so hertz. The hard plastic applicator that is

being used in Fig. 11.8 is applied either with the skin lubricated with oil or talc or with the skin dry. The depth of penetration of the sound waves beneath the skin is up to about 6 cm in soft tissue.

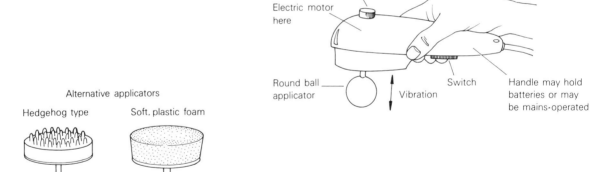

Fig. 11.8 An audiosonic vibration massager

A gyratory or orbital massager

In this apparatus each part of the applicator moves in a small circle driven by an electric motor in the head that rotates or by the rotation of a flexible shaft driven by the motor. A typical frequency for the movement is 40 or 50 Hz.

These units are mostly for body treatments and have quite large heads requiring a heavy-duty motor. Because of the consequent weight of the apparatus, the head is often remotely driven via a flexible shaft with the motor supported on a stand, as shown in Fig. 11.9.

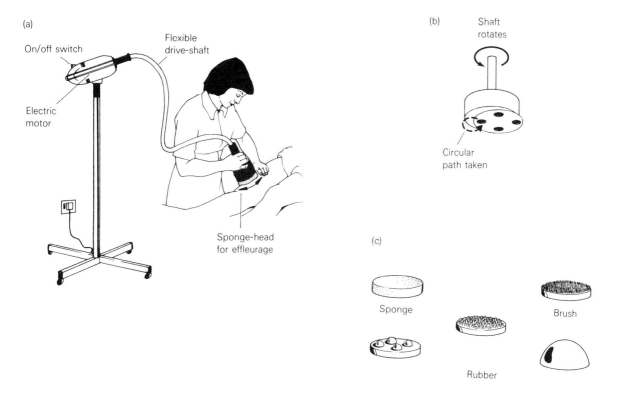

Fig. 11.9 The use of a gyratory massager: (a) massager in use; (b) the gyratory movement of the applicator (see also Fig. 12.14); (c) various applicators

SUMMARY

- Sound travelling in air consists of compressions and rarefactions moving outwards, from the source to the receiver, at high speed.

- The distance from one compression to the next is called the wavelength.

- The number of compressions sent out per second and received per second is called the frequency and is measured in hertz (Hz).

- The speed of sound = wavelength × frequency or $v = \lambda \times f$.

- Ultrasound is of such high frequency that it is inaudible.

- Loudness can be measured in decibels (dB) and is called sound (pressure) level, or in dBA which is a better indication of loudness and is called loudness level.

- Sound can be prevented from reaching an area by use of a barrier (preferably heavy) such as a wall which reflects the sound, and by sound-absorbing materials like acoustic tiles and rock wool.

- About 40 dBA is a suitable loudness level for a clinic, in the absence of clients.

- Audiosonic vibrators produce sound waves in surface body tissues, causing compressions and rarefactions in the tissue. These movements encourage circulation of the blood and lymph and cause a beneficial erythema.

- A gyratory vibrator makes the skin and underlying tissues move in small circles.

1. Distinguish between *audiofrequency* and *ultrasound*.

2. What is the wavelength of a sound of frequency 1000 Hz if the speed of sound in air is 340 m per s?

3. Suggest a suitable maximum loudness level for a clinic when no one is speaking.

4. What is an *audiosonic vibrator?*

5. How does a gyratory massager resemble manual massage?

6. Suggest two ways of preventing noise reaching a particular area of the clinic.

7. How are the frequency and the wavelength of a sound related to its pitch?

12

MAGNETISM AND FURTHER ELECTRICITY

MAGNETS AND ELECTROMAGNETS

Magnets

Steel is iron that has been specially strengthened and it happens to be suitable for making *magnets*. Fig. 12.1 shows a bar magnet and a horseshoe-shaped magnet. (These shapes enable magnets to be made easily.)

Magnets have the ability to attract *iron* and some less well-known metals, but not metals such as aluminium, copper, tin or lead.

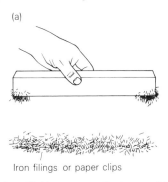

(a)

(b)

Iron filings or paper clips

Fig. 12.1   Picking up small pieces of iron with a magnet: (a) a bar magnet;   (b) a horseshoe-shaped magnet

Magnet poles

Each magnet, as in Fig. 12.1, has two places, close to its ends, where the attraction of iron is most noticeable. These two places are called the *poles* of the magnet.

If a bar magnet is hung on a thread as in Fig. 12.2, it will turn and settle with one pole pointing to the north of the earth and the other pole pointing south. For this reason one pole is called 'north-seeking' (or just 'north pole', N) and the other pole is the 'south-seeking pole' or 'south pole', S.

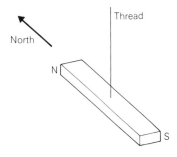

Thread

North

N

S

Fig. 12.2   North-seeking and south-seeking poles

Forces between magnet poles

When two magnets are brought close together and one is free to move we can see that poles exert forces upon each other. A north pole will push another north pole away but will attract a south pole. Similarly an S repels an S and attracts an N pole.

Like poles repel. Unlike poles attract.

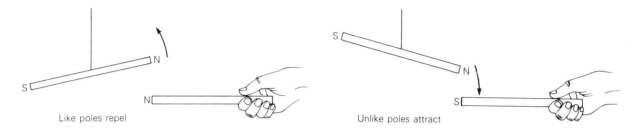

Like poles repel   Unlike poles attract

Fig. 12.3 Forces between poles

Iron near a magnet

When *iron* is placed near to a magnet it becomes magnetised, as shown in Fig. 12.4. The magnetisation is stronger if soft iron is used rather than steel, which is hardened iron.

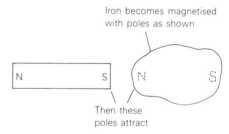

Iron becomes magnetised with poles as shown

Then these poles attract

Fig. 12.4 Iron near a magnet

Soft iron and steel

Soft iron is easily magnetised, for example by putting it near a magnet, but it loses most of this magnetisation as soon as it is moved away from the magnet. *Steel* is more difficult to magnetise, but whatever magnetisation it acquires it will largely keep. We see therefore why steel is more suitable for making a bar magnet than is soft iron.

A compass

A *compass*, which shows us which direction is north, simply consists of a small bar magnet mounted on a pivot so that it is free to rotate. Its north pole end is usually shaped into a point and this end will settle so as to show the way to the north of the earth.

Lines of force of a magnet

These are lines showing the direction in which a compass points (Fig. 12.5) when it is near the magnet.

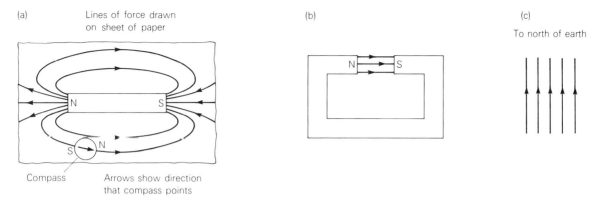

Fig. 12.5   Lines of force: (a) for a bar magnet;   (b) for a magnet with poles facing each other;   (c) with no magnet around

Magnetic fields

The *magnetic field* of a magnet is the area or region around it in which the magnetic effects of the magnet are noticeable. Many small magnets have no easily demonstrated effects at distances of a few centimetres or more from the magnet.

The strength of the field is said to be greatest near to the magnet's poles because it is here that its magnetic effects are strongest. Looking at Fig. 12.5 again, we notice that lines of force are closest together where the field is strongest. In the case of the magnet with poles facing, the strength of the field is the same everywhere between the poles.

The direction of the magnetic field at any place is the direction of the lines of force.

Magnetic fields caused by electric currents

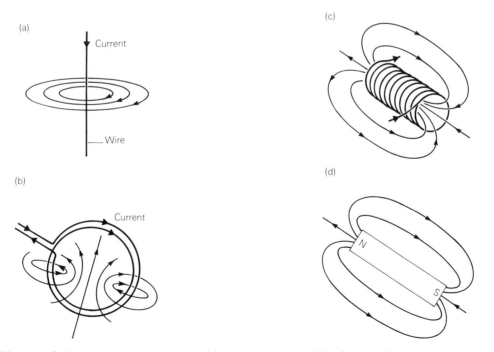

Fig. 12.6   Magnetic fields due to electric currents: (a) a straight wire;   (b) a flat coil of wire;   (c) a long coil (called a solenoid);   (d) a bar magnet for comparison with the solenoid

When a *current* flows through a wire a magnetic field is found to exist around the wire. No magnet is needed, no iron is needed (the wire is likely to be an ordinary copper wire) and it is the wire, or more exactly, the current through it that is behaving as a magnet. Some examples of the lines of force patterns that we get are shown in Fig. 12.6.

Note particularly that the magnetic field *outside* a solenoid is the same as we get with a bar magnet.

The south pole of the solenoid is at the end where the current flows clockwiSe and the north pole where the current flows aNticlockwise (Fig. 12.7).

Fig. 12.7   The poles of a solenoid

Electromagnets

A solenoid carrying an electric current acts as a magnet and can be called an *electromagnet*. However, it is usual to put soft iron inside the solenoid and this becomes magnetised by the solenoid. This magnetised iron 'core' and the solenoid both attract and lift any piece of iron that needs to be moved. The iron core has made the solenoid into a better electromagnet.

Electromagnets have the advantage compared with other magnets that they cease to act as magnets almost as soon as the current is stopped. Consequently the iron that is lifted can be released by switching off the current. Magnets other than electromagnets are therefore called *permanent* magnets.

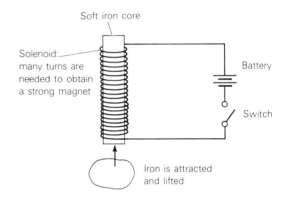

Fig. 12.8   A simple electromagnet

An automatic make-and-break

An electromagnet can be made to provide an automatic stopping and starting of current (usually described as 'make-and-break'). Generators of traditional faradic currents (used for muscle stimulation — see Chapter 4) made use of this method of making and breaking a current. A similar make-and-break is still used in the ordinary electric door-bell.

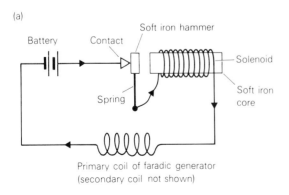

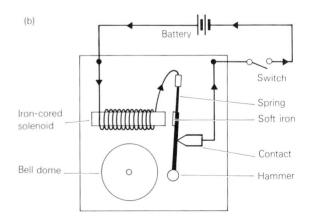

Fig. 12.9   Circuits using an automatic make-and-break: (a) a faradic current generator; (b) an electric bell

In Fig. 12.9, when the switch is closed current flows as indicated by the arrows in the diagrams. The solenoid behaves as a magnet and pulls the soft iron on the hammer. As the hammer moves the electrical contact with the hammer is broken and so the current is stopped. A moment later the spring will have brought the hammer back, the contact, and so the current, being restored. The solenoid will again pull the hammer and the process repeats as long as the switch is kept closed. In this way the hammer repeatedly hits the bell in Fig. 12.9(b) and in Fig. 12.9(a) the current flowing through the primary coil of the faradic machine is repeatedly started and stopped as required.

ELECTRIC MOTORS

Force on a current-carrying wire

A current behaves as a magnet so that we are not surprised to find that a current and a magnet will exert a force on each other (Fig. 12.10).

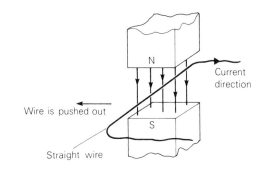

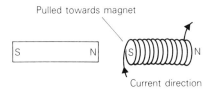

Fig. 12.10   The force on a current-carrying wire

When the current flows in a straight wire at right angles to the lines of force of a magnet the direction of the force on the wire can be deduced from the *left-hand rule*. The first and second finger on the left hand are made to point in the directions of the magnetic field lines and the electric current respectively. The thumb is made to stand out (at right angles to the other two fingers) and its direction is then the direction of the force (and possibly movement). Fig. 12.11 should make this clear.

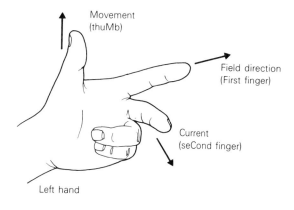

Fig. 12.11   The left-hand rule

A simple electric motor, DC-operated

If a current flows in a flat coil fixed to an axle so that it can rotate between the poles of a magnet, as shown in Fig. 12.12 (a), overleaf, then, by use of the left-hand rule, we realise that the side A of the coil will be pushed down and the side B forced up. The coil rotates and turns the axle.

Before side A reaches the right-hand side the current must be reversed if the rotation is to continue. This reversal can be obtained automatically by use of a *commutator* as seen in Fig. 12.12(b).*

(a)

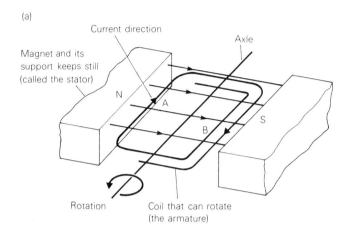

(b)

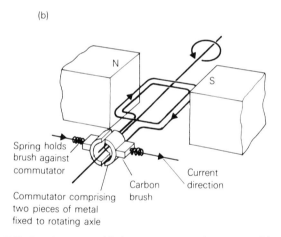

Fig. 12.12   The DC electric motor: (a) the armature and stator;   (b) the use of a commutator

---

AC electric motors
If an alternating current is used for the armature of a motor like the one above there is no need for a commutator because the current reverses (e.g. at 50 Hz) but the motor will work at only one speed, namely 50 Hz. A fixed-speed motor like this is called a *synchronous motor*. A difficulty with the synchronous type of motor is that it needs to have another motor built in or some other device to get it up to the required speed. In practice synchronous motors differ considerably from the construction suggested and the speed need not be 50 Hz.

An alternative type of AC motor, which is more common in fact, is the *induction motor*. The principle on which it works is mentioned later in this chapter.

The use of an electromagnet in an electric motor
Instead of using a permanent magnet in a DC electric motor an electromagnet can be used, the current for it being provided by the voltage supply that provides the current through the armature.

If the motor of Fig. 12.12 had an electromagnet and the voltage supply were AC then not only the armature current would be alternating but so would the magnet field direction. As a result we find by the left-hand rule that the armature will rotate in a constant direction as required. The motor could still be used with a DC supply if desired. A motor of this kind is known as a *universal motor*.

Electric motors in the salon

Already we have met electric motors in vacuum pumps, fan heaters and hair-dryers, ventilation systems and vibration massage apparatus. Fig. 12.13 shows some other motor-driven items that might be found in a salon or health and beauty clinic.

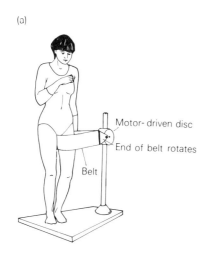

(a)

Motor-driven disc
End of belt rotates
Belt

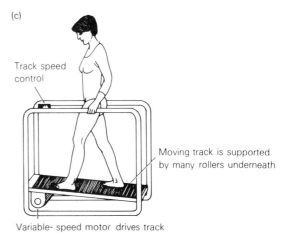

(c)

Track speed control

Moving track is supported by many rollers underneath

Variable-speed motor drives track

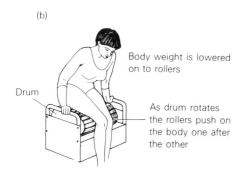

(b)

Drum

Body weight is lowered on to rollers

As drum rotates the rollers push on the body one after the other

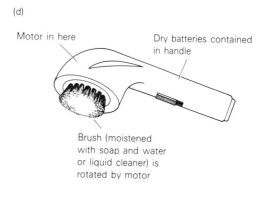

(d)

Motor in here

Dry batteries contained in handle

Brush (moistened with soap and water or liquid cleaner) is rotated by motor

Fig. 12.13   Motorised apparatus: (a) a belt massager;   (b) a roller massager;   (c) a motorised jogging machine; (d) a facial brush

The electrical working of vibration massagers

How can electricity create the in-and-out movements and the gyratory movements of vibration massage devices?

One answer is that the rotation of an electric motor can be used. For example, a gyratory movement can be obtained as suggested in Fig. 12.14 (a), overleaf.

Another answer is that a vibration can be produced by passing an alternating current through a coil which is placed between the poles of a magnet in the manner shown in Fig. 12.14(b). Using the left-hand rule, we find that the coil moves forward when the current flows clockwise in the diagram, and then it moves back when the current direction is reversed. This vibrating coil arrangement is suitable for an audiosonic vibrator. (It is also used in loudspeakers, the cone that vibrates to produce sound being fixed to the front of the coil.)

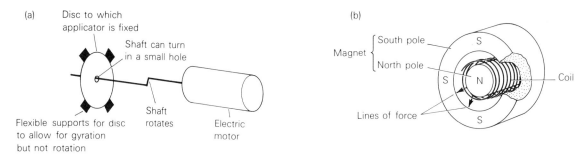

(a)

Disc to which applicator is fixed

Shaft can turn in a small hole

Flexible supports for disc to allow for gyration but not rotation

Shaft rotates

Electric motor

(b)

Magnet { South pole / North pole

Coil

Lines of force

Fig. 12.14   Using electricity for vibration: (a) a gyratory movement;   (b) an in-and-out movement

ELECTRICITY GENERATORS AND TRANSFORMERS

Electromagnetic induction

This is the term used when a voltage is produced by using a magnet instead of a voltaic cell (or a battery). If we obtain a voltage by electromagnetic induction it will, of course, produce a current if there is a complete circuit.

Fig. 12.15 shows how a voltage and current can be produced in a coil by electromagnetic induction.

The magnet must be made to move so that its lines of force are cut through by the turns of the coil and the induction occurs only while the movement is taking place.

The faster the movement, the greater is the induced voltage.

Reversing the direction of the movement, i.e. taking the magnet away instead of bringing it closer, reverses the polarity of the induced voltage.

The same effects are obtained if the coil is moved and the magnet kept still.

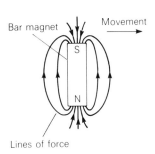

Bar magnet

Movement

Lines of force

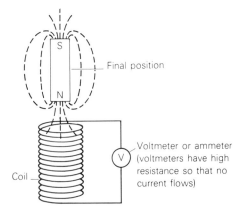

Final position

Coil

Voltmeter or ammeter (voltmeters have high resistance so that no current flows)

Fig. 12.15   Electromagnetic induction

Lenz's law

This law, or rule, named after the scientist Lenz, tells us that the induced voltage polarity and induced current direction is always such that the current will oppose the magnet movement. This is illustrated in Fig. 12.16. (If

the current were to help the movement, we could get a tremendous movement and huge current with negligible effort being needed. This is impossible in view of the law of conservation of energy.)

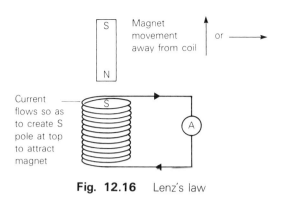

Current
flows so as
to create S
pole at top
to attract
magnet

Fig. 12.16   Lenz's law

Generators

A dynamo is an apparatus that produces a direct current when its armature coil is rotated in the field of a magnet. The apparatus shown in Fig. 12.12 could be used as a dynamo. The current is produced by electromagnetic, induction and, in fact, electromagnetic induction is sometimes called the *dynamo effect*.

A similar apparatus that produces an alternating current is called an *alternator*. Both dynamos and alternators can be called *generators*.

An experiment with two coils

As shown in Fig. 12.17, a current is made to flow in one coil by use of a voltage supply. This coil is called the *primary* and, when current is flowing through it, it behaves as a magnet. The lines of force of this magnet pass through the second coil because this is wound over the primary. Now, if the primary current is stopped, the effect upon the secondary is the same as taking a magnet away from this coil, and so a voltage is produced in the *secondary*. Similarly if the current is started again, the effect will be that of bringing up a magnet and bringing lines of force into the coil. Another induced voltage is created but this time its polarity is opposite to that caused by stopping the current.

It must be emphasised here that we get an induced voltage only while the primary current is changing and that, after the current has been switched on and has settled at a constant value, no induced voltage is produced until the switch-off. The induced voltage at break is usually bigger than at the start because the break is more sudden.

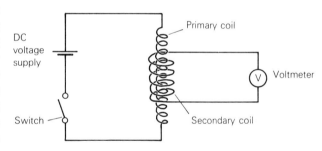

Fig. 12.17   Electromagnetic induction —
the two-coils experiment

If the primary coil is filled with soft iron, then this core of iron becomes magnetised whenever the primary current flows so that the primary is a more effective electro-magnet. The iron therefore makes the experiment work better.

Using a larger number of turns of wire for the secondary coil also results in bigger induced voltages.

Production of faradic currents

Faradic currents are used to exercise muscles and so improve the posture and shape of the body. These currents can be of the true faradic kind that were

produced by electromagnetic induction (and named after Michael Faraday, the scientist who discovered electromagnetic induction) or the modern faradic-type

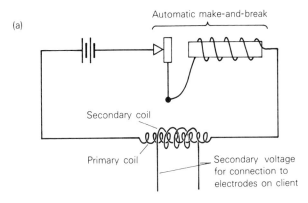

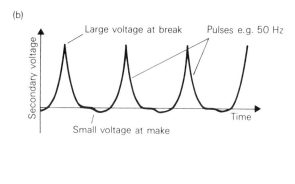

Fig. 12.18   Producing faradic current: (a) a simple circuit;   (b) graph of voltage against time

currents (Chapter 13). The (true) faradic current is obtained using an arrangement of a primary and secondary coil with an automatic make-and-break in the primary circuit, as seen in Fig. 12.18.

Because the induced voltage at the start of the current is small, it is usually of no use and the faradic current can be regarded as being of one polarity only, one of the leads from the secondary being the negative and the other being the positive.

The transformer

If we have a primary and secondary coil arrangement and an iron core different from Fig. 12.17 by having an AC supply, then, because an alternating current is continually changing in size, we do not need a switching on and off or an automatic make-and-break to cause changes in the primary current. The primary current grows and then equally rapidly decreases and we get induced voltages in the secondary coil which also alternate in polarity, as shown in Fig. 12.19.

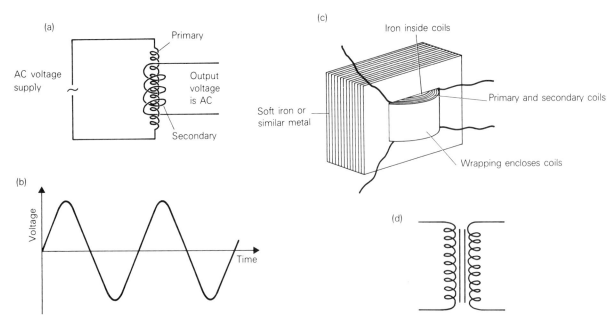

Fig. 12.19   The transformer: (a) the circuit;   (b) the secondary voltage;   (c) the appearance of a typical transformer; (d) the circuit symbol

134

Transformers are extremely useful because the alternating voltage obtained from the secondary can be made greater than the voltage supplied to the primary (although the currents obtained will necessarily be smaller), or the secondary voltage can be smaller than the primary voltage with larger currents being possible.

A transformer can replace an alternating voltage by a bigger one or a smaller one.

The size of the secondary AC voltage can be calculated from the formula:

$$\frac{\text{Secondary voltage}}{\text{Primary voltage}} = \frac{\text{No. of turns on secondary}}{\text{No. of turns on primary}}$$

$$= \textbf{Turns ratio}$$

and the secondary and primary currents are such that:

$$\frac{\text{Secondary current}}{\text{Primary current}} = \frac{1}{\text{Turns ratio}} \text{ or less than this}$$

A transformer that gives a secondary voltage bigger than the voltage supplied to its primary (i.e. the turns ratio is bigger than 1) is called a *step-up transformer*. A *step-down transformer* gives a reduction in voltage.

From what has been said already one should expect a high secondary voltage if the secondary has many turns but the theory of the transformer is too difficult for us to explain these facts here. This is because as soon as alternating currents flow in the secondary they cause some electromagnetic induction in the primary and make the whole picture most confusing.

Transformers will be found inside many pieces of electrical apparatus in the beauty salon changing the mains supply voltage of 240 V to lower values (see Fig. 13.8(b) p. 142) or, for example in the spark-producing high-frequency apparatus (Fig. 13.18, p. 149, increasing the voltage.

DC power packs are devices that are plugged into the mains supply and produce a DC voltage, often 6 V or 9 V, which may be required by equipment intended for battery operation. Such power packs have a step-down transformer inside as well as a rectifier (see Chapter 13).

Two practical points to remember are: (1) that the leads from the secondary should never be shorted for fear that the large secondary current and the consequent large *primary* current may ruin the transformer, perhaps with smoke and fire adding to the disaster, and (2) that when the transformer is off-load (no current being taken from the secondary), then the primary current will automatically be small as will the electricity bill.

The induction motor

A simple motor of this type comprises an electromagnet that is fed with an alternating current and an armature that is simply a drum on which copper bars are fitted and joined to form loops. Currents are induced in the bars by the ever-changing field of the electromagnet and, as a result, the armature rotates.

SUMMARY

- Magnets can attract iron and some less well-known metals but not common metals such as copper, tin, lead or aluminium.

- The attraction is greatest at the poles of the magnet.

- A suspended magnet, such as a compass, settles with always the same pole pointing northwards. This pole is called the north (seeking) pole. The other is the south.

- Like poles repel each other; unlike poles attract.

- Soft iron placed near a magnet becomes magnetised, but it easily loses its magnetisation. Steel is more difficult to magnetise but it retains whatever magnetisation it is given.

- Lines of force show the direction in which a compass needle points.

- An electric current behaves like a magnet, a solenoid carrying a current being just like a bar magnet.

- A solenoid with a soft iron core is a useful electromagnet. When the current stops, it ceases to be a magnet.

- An electromagnet is used as part of an automatic make-and-break in electric bells and in the original kind of faradic machines.

- A wire carrying a current and placed near a magnet experiences a force and use is made of this in electric motors.

- The alternating force on a coil of wire carrying an alternating current (with a magnet present) can be used in an audiosonic vibrator.

- Electric motors are used for vibrators, belt massagers, some jogging machines and for driving pumps in vacuum equipment and fans (for example, in hairdryers).

- Electromagnetic induction is the production of a voltage by moving a wire near a magnet or vice versa.

The induced voltage is of such polarity that any current it causes opposes the cause of the induction, i.e. the movement. Moving a magnet towards or away from a coil produces a voltage in this way.

- An electromagnet (the primary coil) can be used with a secondary coil to produce electromagnetic induction. This principle was used in the old type of faradic apparatus, the current changes in the primary being caused by an automatic make-and-break.

- If the primary coil is supplied with AC, then the secondary induced voltage is AC also and the primary and secondary are acting as a transformer.

- A step-up transformer gives a larger secondary voltage than the primary. A step-down transformer gives a smaller secondary voltage but allows a larger secondary current to be obtained.

EXERCISE 12

1. Which of the following materials is attracted to a magnet:

copper, tin, lead, iron, aluminium, carbon?

2. What are the characteristics of steel as regards being magnetised?

3. Describe a simple electromagnet.

4. Describe an automatic make-and-break. What can it be used for?

5. Name three uses for electric motors in a beauty clinic.

6. What is meant by *electromagnetic induction?*

7. Name the main parts of a typical transformer and explain what it can be used for.

13

ELECTRONICS

THE AMAZING ELECTRON

Electrons are amazing because they are electrically charged and they weigh very little. As a result we can make them move very rapidly and stop and start them extremely quickly. The science and technology that makes use of this is called *electronics*. The discovery of diodes and transistors has made electronics a huge success story and it has given us, among other things, transistor radios, record players, tape recorders, television, calculators and computers. All this because of what we can do with the electron!

In this Chapter electronics is explained so that the working of faradic, high-frequency, diathermy and other equipment may be understood.

CONVERSION OF AC TO DC

Rectifiers

One of the most useful devices for controlling electron movement is a *rectifier*. A rectifier has the property that it *lets electrons flow through it in one direction only*, rather like a one-way street.*

Rectifiers may be made out of metal and silicon (a 'metal rectifier') or of silicon treated so as to create regions of so-called p and n type silicon within the diode. The appearance and actual size of some small rectifiers is shown in Fig. 13.1.

A simple rectifier is called a *diode* (meaning 'two electrodes'). This name is used partly for historic reasons but it does fit the fact that the diode has two leads whereas a transistor has three or more.

In circuit diagrams we use the symbol shown in Fig. 13.1(c) to represent a silicon diode or metal rectifier. The direction in which the symbol points is the direction in which the rectifier allows current to flow.

Some diodes when conducting give out light and are called *Light Emitting Diodes* or LEDs.

Fig. 13.1   Examples of small rectifiers: (a) the appearance of silicon diodes;   (b) the appearance of a metal rectifier;   (c) the circuit symbol for a rectifier

*In fact electrons can flow in the other direction, but the resistance is very large. This small current in the reverse direction is called *leakage current*.

Rectification

This means obtaining a direct current from an alternating voltage. A rectifier is used. A simple *rectification circuit* is shown in Fig. 13.2. It is called a *half-wave* rectifier circuit because of the shape of the current versus time graph.

The current obtained with half-wave rectification is not ordinary DC because it is not a steady flow of current. This can be improved, as shown later in this chapter, by use of a *capacitor*.

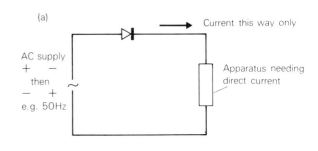

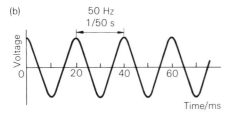

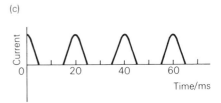

Fig. 13.2   A half-wave rectifier circuit: (a) the circuit;   (b) the supply voltage;   (c) the current

A full-wave rectifier

Looking at Fig. 13.3 we see that there is a current flowing through the load, always in the same direction, even when the supply voltage reverses. This is a more successful rectifier than the half-wave one.

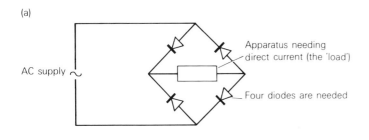

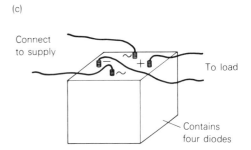

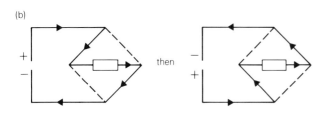

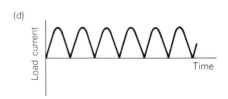

Fig. 13.3   A full-wave rectifier: (a) the circuit;   (b) the current flow;   (c) a ready-made, full-wave rectifier;   (d) the load current

⊐ LIGHT AND ULTRA-VIOLET RADIATION BY USE OF ELECTRICITY ⊏

——————————————— Conduction in gases ———————————————

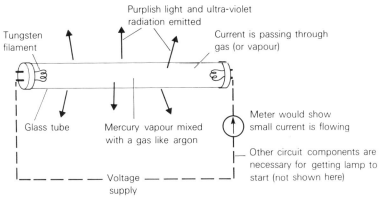

Fig. 13.4   Conduction in a gas (or vapour) — a low-pressure mercury vapour lamp

Gases are usually very good electrical insulators, but if the pressure of a gas is reduced and a sufficiently high voltage is used, conduction can occur and it is then accompanied by release of electromagnetic radiation which is often visible light.

Examples of conduction in a gas can be seen in the form of neon display lights, the modern yellow (sodium vapour) street lamps, fluorescent tube lamps used in the home and workplace, and ultra-violet lamps now common in many disco halls. Ultra-violet lamps are also used in beauty clinics for sun-tanning and are found in ozone steamers and in ultra-violet sterilisers.

Fig. 13.4 shows that, once the lamp has 'got started', a current flows, passing through the gas from one tungsten electrode to the other and both light and ultra-violet radiation are emitted from the tube.*

The colour of the light released depends very much upon the type of gas used in the tube. Sodium vapour readily produces yellow light, neon produces a useful amount of red light, white mercury vapour produces mostly ultra-violet radiation.

Fig. 13.4 shows that coils of tungsten wire are used for the lamp's electrodes. Use is made of these when the lamp is first switched on. A current is passed through the coils to heat them and they in turn vaporise some of the liquid mercury in the tube. After this the voltage is connected, as shown in the diagram, to make the gas conduct and glow.

Lamps can also be made that work with the gas mixture at high pressure, for example, the high-pressure mercury vapour lamp (HPMV lamp).

Conduction is more difficult in a gas at high pressure, as mentioned already, but a small tube is used (made of quartz to withstand the very high temperature that will be reached) and special design overcomes the problem of starting the conduction. Once started, this lamp, in spite of its size, gives out a great deal of light as well as ultra-violet radiation. These lamps are used in some sun-tanning systems.

---

*If the gas mixture in this lamp were at atmospheric pressure, only a very small current could be obtained because only a few electrons here and there in the gas would be free rather than held in the gas molecules. These free electrons would move towards the positive electrode and so a tiny current would be obtained. The effect of reducing the pressure of the gas is to space out the molecules giving the free electrons moving towards the anode sufficient room to get up to high speeds. If they then bump into gas molecules at these high speeds, they have enough kinetic energy to knock electrons out of the molecules hit, and a vast increase in the number of free electrons results. In this way an appreciable current is possible. Perhaps you think that the current will increase indefinitely? Well, it would and this could make the circuit overheat. A resistor in series with the lamp can prevent this, but a choke (see p. 146) is better.

The release of light, etc. occurs when an electron hits an atom fairly hard, moving an electron in the atom and giving it some kinetic energy but not enough to set it free. The atom is seen to be excited. The electron is then attracted back by the atom's nucleus to its normal position in the atom and the energy that was given to the electron is released from the atom as light or other electromagnetic radiation.

ELECTRONICS IN ELECTROTHERAPY EQUIPMENT

The transistor

The appearance of one kind of small *transistor* and its circuit symbol are shown in Figs. 13.5 (a) and (b).

The type of transistor shown in the diagram is called an *FET* and the three terminals of an FET are called the *source*, the *drain* and the *gate*. It is described as *n-channel* because the channel through which the current flows in the transistor is made of n-type silicon (silicon made conducting by adding some free electrons).

Transistors are used a great deal in *amplifiers*. A simple amplifier circuit is shown in Fig. 13.5(c).*

Amplifiers are useful, for example, in radios, televisions, intercoms, etc. In beauty therapy equipment they are found in faradic apparatus (modern type) and in epilation equipment of the diathermy type. Small currents or voltages from thermometers can also be amplified. (Transformers might be used to amplify some voltages but then the current obtainable is always reduced as already explained.)

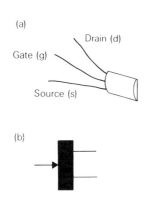

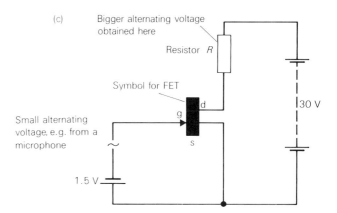

Fig. 13.5 A transistor and its use, a small n-channel FET: (a) appearance of an FET; (b) the circuit symbol for an n-channel FET; (c) a simple amplifier circuit

Capacitance and capacitors

Capacitance means 'ability to store electric charge'. Any two conductors with an insulating layer between them will have this ability or capacitance, to some extent. The capacitance can be usefully large if the conductors are of large area and separated by only a thin layer of insulator. The insulating material can be called the *dielectric* (and

the type of dielectric does in fact affect the capacitance). Fig. 13.6 shows some capacitors and the circuit symbol.

A combination of two conductors with dielectric between them that has been constructed in order to

---

*When the transistor is connected to a voltage supply, like the 30 V battery shown in the figure, we expect electrons to flow in at the source and out through the drain. As this current (*I*) flows through the resistor (*R*) we get a p.d. of $V = I \times R$ across the resistor. If the gate terminal is made negative compared to the source, using perhaps 1½ V as shown, this has the effect of reducing the current flowing through the transistor, just as if a gate has been closed a little through which the current has to pass. Now if a microphone produces a small alternating voltage, this will make the gate terminal alternately more negative and then less negative so that the gate is alternately closed more then less. Consequently the size of the current through *R* falls and rises as does the p.d. across *R*. These fluctuations of voltage are larger than the voltage fluctuations that were provided by the microphone so that we have a voltage amplifier.

provide a useful capacitance is called a *capacitor*. One common type of capacitor is made of two thin sheets of metal, often of very large area, with a very thin layer of plastic or waxed paper as the dielectric between them.

It is difficult to store much charge in a single piece of metal because the first charge put in repels any further charge that we try to add. The success with which charge is stored in a capacitor is due to the + and − charges being close together and attracting each other. The attraction helps to hold the charges in the capacitor.

The capacitance of a capacitor is measured in farads, the abbreviation for farad being F. The microfarad (μF) is usually a more convenient unit.

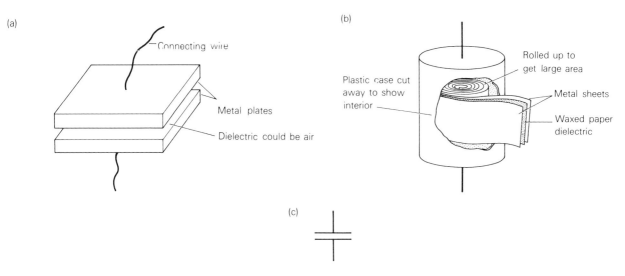

Fig. 13.6   Capacitors: (a) a simple capacitor;   (b) appearance of a waxed paper capacitor;   (c) circuit symbol for a capacitor

Charging of a capacitor

The charging of a simple capacitor is illustrated in Fig. 13.7(a).

With a DC supply the current soon stops. The capacitor is then said to be *fully charged*.

The current has stopped because the capacitor's conductors are now as + and − (at their outer surfaces) as are the terminals of the supply battery.

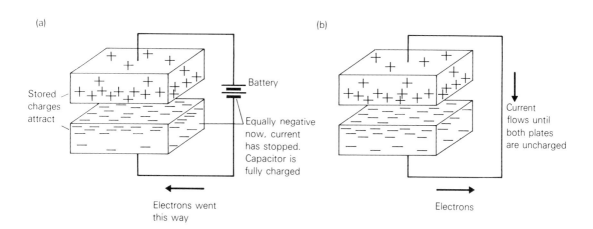

Fig. 13.7   Charging and discharging of a simple capacitor: (a) charging;   (b) discharging

Discharging a capacitor

If the two conductors of a charged capacitor are joined so as to form a circuit but without the charging battery present, then electrons will move from the negative conductor round to the positive conductor until both conductors are uncharged, i.e. the capacitor discharges. During discharge the current flow is opposite in direction to the current during charging (Fig. 13.7(b)).

Use of a capacitor to smooth the output from a rectifier

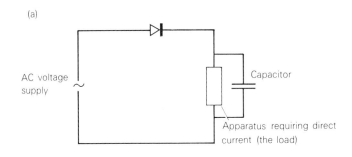

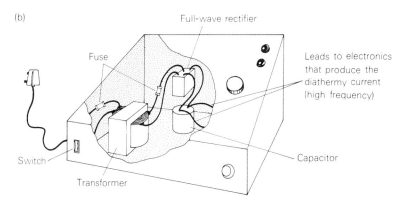

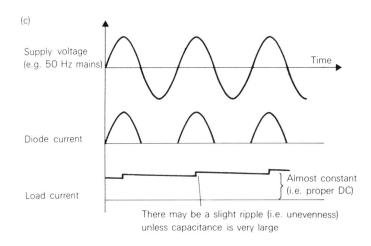

Fig. 13.8   Use of a capacitor to smooth the output from a rectifier: (a) circuit;   (b) as seen in a therapy unit e.g. diathermy epilation unit;   (c) a comparison of load current with the diode current

A capacitor may be used to *smooth* the output from a rectifier. Fig. 13.2 (a) (on p. 138) shows a circuit without a rectifier. When a rectifier is added (Fig. 13.8) the graph of current against time is almost a straight line. The current has been smoothed. The capacitor smooths the current in the load by storing some charge when the rectifier diode is letting charge through, and then discharging this through the load when the load is not able to receive any through the diode.

A capacitor in an AC circuit

Whereas a capacitor soon brings current to zero with a DC voltage supply, we get alternating current flowing continually when the supply is alternating (AC).

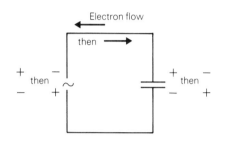

(Current flow in opposite direction to electron flow, of course)

Fig. 13.9   A capacitor in an AC circuit

Electrons flow through the wires (Fig. 13.9) and the capacitor charges just as with a DC supply, but then the polarity of the voltage supply changes so that the capacitor discharges and recharges with the electrons flowing in the opposite direction. Thus we have alternating current in the circuit as the discharging–recharging cycle repeats continually.

A capacitor allows alternating current to flow as if the current is flowing 'through' the capacitor. The flow will be easiest if the capacitance is large and if the frequency is *high* (time for charging short).

Capacitance of the skin

The skin is not a very good conductor of electricity even when it *is* moist, and discomfort is first felt in the skin if we attempt to pass too large a current through any part of the body. However, if we think of an electrode placed on the skin and remember that the tissues beneath the skin conduct quite well, we realise that we have a capacitor with the skin acting as the dielectric (Fig. 13.10). Advantage is taken of this fact when so-called *interferential currents* are used instead of the usual faradic currents.

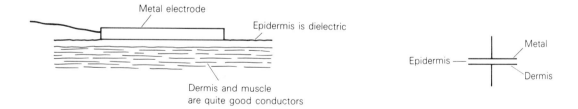

Fig. 13.10   The capacitor formed by the skin acting as a dielectric

Interferential currents (Beat currents)

Interferential current therapy uses currents at medium frequencies because at these frequencies the currents can pass easily 'through' the skin since it is acting as a capacitor. There is little or no heating or discomfort in the skin.

Two equal sinusoidal voltages of slightly different frequencies, f_1 and f_2, are connected as shown in Fig. 13.11 to two pairs of electrodes.*

The current up and down the leg becomes strong and weak alternately with a frequency equal to the difference between f_1 and f_2, which would be 10 Hz if the medium frequencies were (say) 4000 and 3990 Hz. Now, muscles cannot respond to frequencies above 100 Hz, but they can respond to the lower frequency of 10 Hz at which the current strengthens or beats. The *beats* are

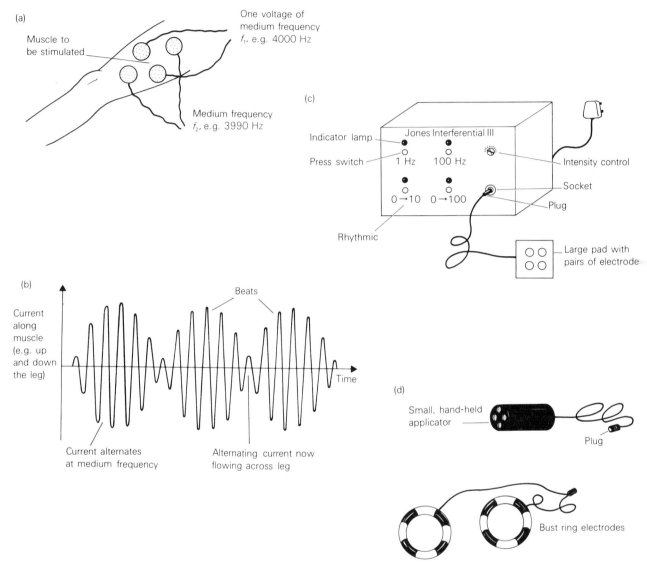

Fig. 13.11 Interferential current therapy: (a) the positions of electrodes; (b) the current along the muscle; (c) a typical interferential supply unit; (d) examples of other pads

*If the second voltage only were used, there would be an alternating current in the leg, that is, up the leg then down alternately. With the first voltage only there would be just an alternating current flowing across the leg. If both voltages are applied, then the current at the middle between the four electrodes can flow up and down, across or in other directions and, because of the small difference of frequencies the current at the middle flows in all the directions in turn. If we suppose that a current is required at the middle to flow up and down the leg because this is the best direction to stimulate a certain muscle, then we need to know that this current will vary as shown in Fig. 13.11(b). At one stage we have a strong alternating current up and down the leg at about 4000 Hz, but later the current in this direction is zero, there being no flow up and down, only across, the leg.

therefore the result of the combined effect of the two voltages and the intermingling of the two currents which is described as *interference* between the two currents or voltages.

Because the direction of the alternating current flow rotates, a muscle will regularly receive a current in a suitable direction.

If one of the two frequencies is kept at 4000 Hz while the other one is slowly decreased from 4000 Hz to 3900 Hz, then the beat frequency changes from zero up to 100 Hz. This has the advantage, presumably, of at least providing the most suitable frequency at some stage and it stops the muscle from acclimatising to one frequency and responding less. This sweeping through a range of frequencies is usually called *rhythmic*. For strong contractions of muscles, frequencies of 1 to 10 Hz may be best. Higher frequencies, particularly about 50 Hz, cause continuous contraction (tetanus), and frequencies approaching 100 Hz or more, depending upon the current strength, produce no contractions but are used to stimulate the circulation of the blood and lymph.

Coils

Suppose that a direct current is flowing in a coil of copper wire so that a magnetic field is present with lines of force passing through the middle of the coil. If this current is stopped or changed suddenly, there will be an induced voltage produced in the coil. This effect is called *self-induction*.

A coil which is designed to make use of self-induction (rather than designed to provide resistance) is called an *inductor* and it is usually made of copper wire, very often wound on a soft iron core to increase the self-induction. (See Fig. 13.12.)

A coil's ability to produce this effect is called its *inductance* and this will be large if the coil has many turns, especially if it is wound on soft iron.

The unit used for measuring inductance is the *henry* (H).

(a)

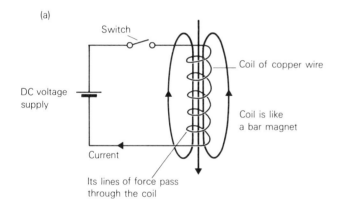

(b)

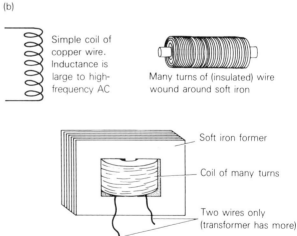

Fig. 13.12   Self-induction: (a) its cause;   (b) the appearance of some inductors

A hazard due to self-induction

A sudden change of current in a coil of large inductance (caused by switching off the supply) can sometimes produce a dangerously high and often unexpected voltage. So make sure that you never touch terminals of unfamiliar equipment (even if it normally handles only low voltages) when switching off.

This fact is emphasised by the demonstration experiment illustrated in Fig. 13.13 (overleaf).

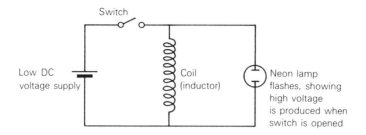

Fig. 13.13   Very high voltages can arise from self-induction

Inductance in an AC circuit

An *inductor* in a circuit (often simply called a *coil*) opposes the flow of alternating current.*

Opposition to alternating current flow is called *impedance*. The impedance caused by the presence of a coil is highest for high-frequency currents.

An inductor is used, for example, to limit the size of current that flows when a fluorescent lamp or other lamp making use of gas conduction is operated on AC supply. When used for this purpose it is called a *choke*.

Impedance

As already stated, this is the opposition to alternating current flow in a circuit. It may be caused by:

(a)   resistance, just as with direct currents,

(b)   inductance, as explained above,

(c)   presence of a capacitor of insufficient capacitance to make flow easy.

Inductance (*L*) and capacitance (*C*) in series with an AC supply

If a circuit of this description is supplied with an alternating voltage whose frequency can be varied, the current (peak or RMS) obtained is small at all frequencies except at one frequency called the *resonant frequency* of the circuit (Fig. 13.14). The value of this frequency is decided by the values of *C* and *L*. [†]

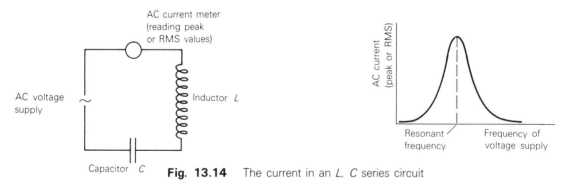

Fig. 13.14   The current in an *L, C* series circuit

---

*This is because, when an alternating current flows through a coil, induced voltages are produced as the current rises and falls. By Lenz's law these voltages must oppose what causes them, i.e. the current changes, so that alternation of current is difficult.

[†] The current is large, the impedance small, at the resonant frequency. This is because at this frequency the induced voltage in the coil is exactly equal in size and opposite in polarity to the voltage across the capacitor at all times so that the combined effect of these two voltages is zero. Therefore there is no limiting effect upon the flow of alternating current (except for any resistance present due, for example, to the use of *thin* copper wire for the coil).

L and C in parallel

This situation is illustrated in Fig. 13.15. In this case the current obtained is high except at one frequency, the resonant frequency of the circuit.

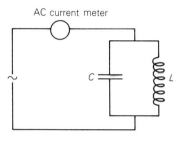 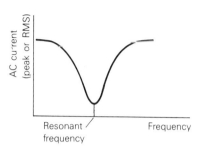

Fig. 13.15   The current in an *L, C* parallel circuit

Spontaneous oscillations

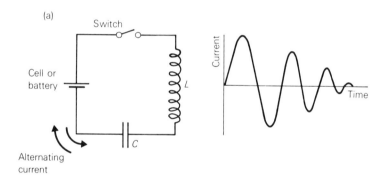

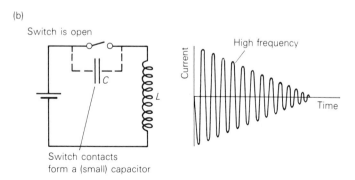

Fig. 13.16   The spontaneous oscillation of current: (a) when switching on;   (b) when switching off

A sudden change of the supply voltage to an L, C circuit can cause a temporary alternating current or oscillation of current (Fig. 13.16).*

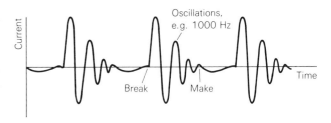

Spontaneous oscillations of current occurred in some designs of (original) faradic machines, the capacitance concerned being largely due to a capacitor fitted between the hammer and the contact. The resulting faradic current produced is as shown in Fig. 13.17.

The traditional 'high-frequency' apparatus used by hairdressing specialists and by beauty therapists makes use of spontaneous oscillation, but obtains the change in voltage by use of a *spark gap*, as explained below.

Fig. 13.17   A faradic current with spontaneous oscillations

The high-frequency therapy apparatus

This is shown in Fig. 13.18. The spark gap consists of two pieces of metal close together with an air gap between them. The air is at atmospheric pressure but the gap is small so that conduction will occur through this air if the voltage across the gap becomes high enough. The current through the air then flows as a spark (not a glow that occurs if the air pressure is low). The mains AC supply voltage is stepped up by the first transformer and the resulting large voltage as it alternates repeatedly reaches a size that causes a spark and hence a drop in the gap resistance. The voltage supply to the L, C circuit is thereby reduced suddenly and oscillation occurs.

The noise that is heard from this equipment is the sound of the sparks in the spark gap.

The graph of the output voltage is sketched in Fig. 13.18(c), (overleaf). The frequency of the output voltage is about a million hertz and the (peak) voltage is some thousands of volts.

The current passing between the electrode and the skin consists of numerous small sparks. The current obtainable is limited by the impedance of the apparatus so that a current of lethal size is not possible.

Since a spark is a brief current through a gas it is not surprising that light is released. Also, the *oxygen ions* (atoms from which electrons have been knocked out), formed in the air when it conducts, can join up to form *ozone gas*, which can kill germs in small concentration (and is poisonous in large concentration). These effects occur both in the spark gap and in the space separating the electrode from the skin.

It may be surprising that there is not a complete circuit of conductor for the therapy current to flow in. The reason for this is that the very high frequency of the current enables it to pass 'through' air spaces and furniture by capacitor action along the route shown by the dashed line in the figure (see page 143). It is also capacitor action that explains how current flows through the glass walls of glass electrodes.

Fig. 13.19 shows how this apparatus is used in the indirect method.

Both the direct and indirect methods produce warmth and encourage blood and lymph flow. The indirect method allows simultaneous massage, while the direct method, in which the current is accompanied by sparking between the electrode and the skin, is very

*This alternating current is explained as follows. If the switch in Fig. 13.16 (a) is suddenly closed, the voltage supply produces a current and C begins to charge up. Then the current falls as the capacitor becomes more charged. This decrease of current causes a voltage to be induced in L, and this continues the current so that C becomes charged still more. Subsequently C begins to discharge because the supply voltage is not enough on its own to keep C so much charged up. Then this discharge current will fall and cause an induced voltage which continues this current even when C is discharged so that C recharges with opposite polarity. As this current falls we get an induced voltage again, C overcharges again and the whole process repeats. Thus an alternating current occurs. It will, however, fade away because the electron movement is opposed by resistance (the circuit will unavoidably have some resistance) so that each flow of electrons is weaker than the last. The frequency of the oscillation is decided by the values of L and C and is the resonant frequency of the circuit, exactly the same as for the L, C series circuit.

In a similar way, if a circuit contains a considerable inductance and the circuit is suddenly broken (by opening the switch in Fig. 13.16(b)), then the sudden drop of voltage causes an oscillation. In the Figure the metal contacts in the switch form the capacitor and the value of the capacitance C is small, resulting in a high resonant frequency at which the current alternates.

(a)

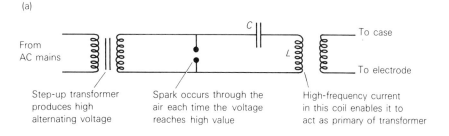

From
AC mains

C

L

To case

To electrode

Step-up transformer
produces high
alternating voltage

Spark occurs through the
air each time the voltage
reaches high value

High-frequency current
in this coil enables it to
act as primary of transformer

(b)

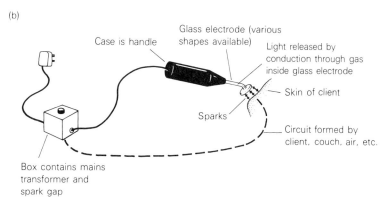

Case is handle

Glass electrode (various
shapes available)

Light released by
conduction through gas
inside glass electrode

Skin of client

Sparks

Circuit formed by
client, couch, air, etc.

Box contains mains
transformer and
spark gap

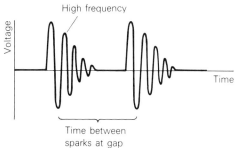

Voltage

High frequency

Time

Time between
sparks at gap

Fig. 13.18   High-frequency apparatus: (a) the circuit (slightly simplified);   (b) the external appearance

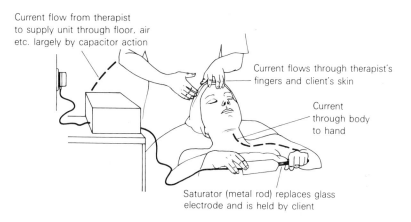

Current flow from therapist
to supply unit through floor, air
etc. largely by capacitor action

Current flows through therapist's
fingers and client's skin

Current
through body
to hand

Saturator (metal rod) replaces glass
electrode and is held by client

Fig. 13.19   The indirect high-frequency method

149

stimulating and tends to dry the skin, and the ozone produced may have a beneficial antibacterial action.

In both treatments *it is essential that no sudden break in* the circuit should occur because of the resulting dis*comfort*. Contact with the client must be maintained until the voltage has been slowly reduced to zero.

A transistorised oscillator

In Fig. 13.20(a), when the switch is closed a current flows through the transistor and in the *L, C* circuit. Temporary oscillation is expected in the *L, C* circuit, at the resonant frequency, but as the alternating current flows through *L* it causes an induced voltage, by transformer action, in the feedback coil. This voltage fed back into the gate circuit causes a change in the transistor current and this change keeps the oscillation going, and so on, repeatedly.

The output voltage from a simple transistorised oscillator is sinusoidal. The frequency can be varied by using a capacitor (*C*) whose capacitance is variable, as shown in Fig. 13.20(b).

An interferential current apparatus (Fig. 13.11, p. 144) provides two alternating, sinusoidal voltages and contains two oscillators although the circuits are not exactly the same as the one described here.

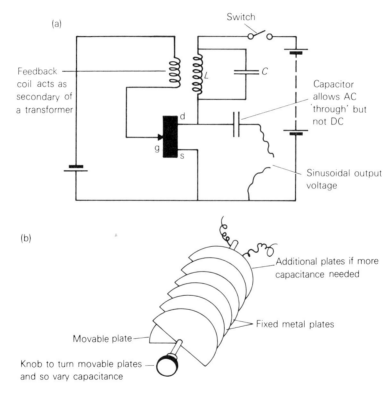

Fig. 13.20 A transistorised oscillator: (a) the circuit diagram; (b) a variable capacitor

A diathermy apparatus for epilation

For *epilation* by *diathermy* a current of very high frequency is used, like the high-frequency sparking apparatus but not at such a high voltage that sparking occurs. Because of the very high frequency, the only effect upon the body is heating. Fig. 13.21 shows the way in which it is used.

The intensity control can be a variable capacitor (see Fig. 13.20(b)) whose impedance will increase as the capacitance is reduced.

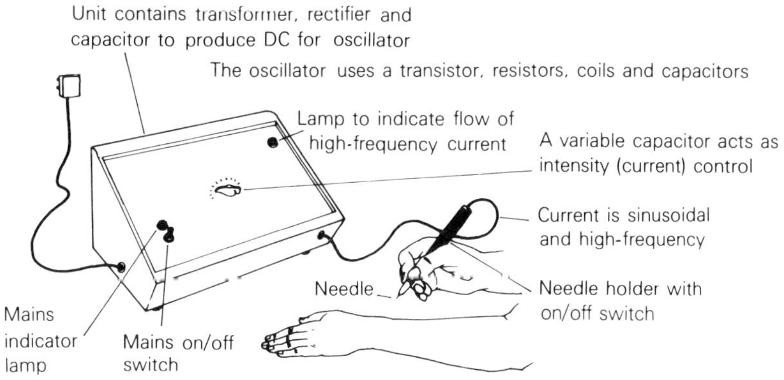

Unit contains transformer, rectifier and
capacitor to produce DC for oscillator

The oscillator uses a transistor, resistors, coils and capacitors

Lamp to indicate flow of
high-frequency current

A variable capacitor acts as
intensity (current) control

Current is sinusoidal
and high-frequency

Needle

Needle holder with
on/off switch

Mains
indicator
lamp

Mains on/off
switch

Fig. 13.21   Epilation using diathermy

The heating is very intense around the needle, especially close to its point (Fig. 13.22), and the base of the hair root and part of the follicle are destroyed literally by cooking.

The diathermy and galvanic methods of epilation can be combined, the two effects occurring around the same needle. The technique is known as the *blend*.

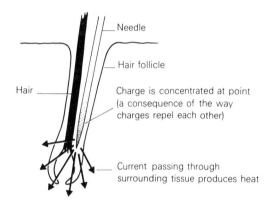

Needle

Hair follicle

Hair

Charge is concentrated at point
(a consequence of the way
charges repel each other)

Current passing through
surrounding tissue produces heat

Fig. 13.22   Heating is greatest close to the point of the epilation needle

Modern faradic units

The current required from a modern *faradic unit* is a series of pulses (often of increasing size to produce a 'surge') followed by a period of rest before the pulses start again (Fig. 4.23 p. 37). An electronic unit for producing such a faradic-type current is shown in Fig. 13.23. It contains a special kind of oscillator.

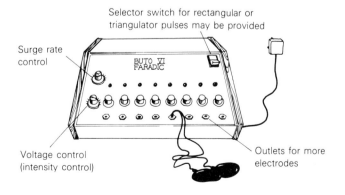

Selector switch for rectangular or
triangulator pulses may be provided

Surge rate
control

BUTO VI
FARADIC

Voltage control
(intensity control)

Outlets for more
electrodes

Fig. 13.23   A supply unit for producing faradic-type current

Integrated circuits (silicon chips)

An *integrated circuit* is a block or 'chip' of silicon, parts of which have been made into minute transistors, diodes, resistors and capacitors with strips of metal coated on to the surface to connect these components together to form the required circuit. The chip is enclosed in a plastic case and its leads (pins) connected to terminals as shown in Fig. 13.24.

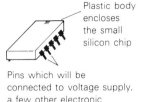

Plastic body encloses the small silicon chip

Pins which will be connected to voltage supply, a few other electronic components, fuses or lamps and suitable output terminals, etc.

Fig. 13.24   The typical appearance and size of an integrated circuit

SUMMARY

- A diode allows electric current to flow in one direction only.

- A rectifier gives a direct current with an alternating voltage supply.

- A full-wave rectifier is more effective than a half-wave rectifier but uses four diodes.

- An electric current can be made to flow through a gas or vapour, especially if the gas pressure is reduced.

- When current flows through a gas or vapour electromagnetic radiation is emitted.

- A lamp that has a current flowing through a gas needs a resistor or a choke to limit the current.

- Transistors are important in amplifiers and similar circuits.

- Capacitors are devices that store electric charge, + charge on one plate and − charge on the other plate. The insulator between the plates is called the dielectric.

- Capacitors are used for smoothing the output from rectifiers.

- An alternating current can flow 'through' a capacitor. This effect can be called *capacitor action.*

- The flow of high-frequency current 'through' the glass walls of glass electrodes is due to capacitor action.

- Medium-frequency currents used in the interferential or beat-current technique pass 'through' the skin easily due to capacitor action.

- Coils oppose the flow of AC due to voltages induced in the coils.

- A coil and capacitor combination can be used to produce an oscillatory current, for example, in high-frequency (sparking) apparatus.

- Chokes are coils for limiting the size of an alternating current.

- Electric circuits employing coils and capacitors and transistors are useful for producing sinusoidal currents for diathermy epilation. Other electronic circuits are used for modern faradic-type currents.

EXERCISE 13

1. What is the function of a rectifier in an electrical circuit? (CGLI)

2. What is meant by *impedance?*

3. What are *interferential currents?*

4. Distinguish between a diode and a transistor such as an FET.

5. What is a capacitor? Name the essential parts of a capacitor.

6. How does a capacitor differ in its effects in an AC circuit compared with a DC circuit?

7. Give one example of the use of a capacitor in a piece of beauty therapy equipment.

8. What special property makes a coil and capacitor combination useful in a sinusoidal oscillator?

9. Explain why a variable capacitor can control the intensity of high-frequency current.

14

LIGHT — MIRRORS AND LENSES

MIRRORS

Plane mirrors

If there were no mirrors would one be so conscious and concerned about one's appearance and beauty? Perhaps we take mirrors too much for granted.

How does a flat (*plane*) mirror work? First we observe that light usually travels out in straight lines from its source, for example, from a lamp. It streams out rather like the steam spirting out in all directions from the saucepan in Fig. 14.1(a), but light's speed of travel is enormous (300 million metres per second). In Fig. 14.1(b) a narrow strip of light (called a *ray*), after travelling in a perfectly straight line, is changed in direction by the mirror because the light 'bounces off the mirror', i.e. it is *reflected*.

(a)

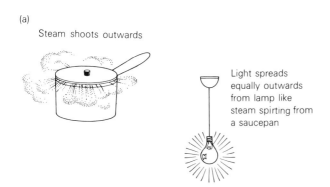

Steam shoots outwards

Light spreads equally outwards from lamp like steam spirting from a saucepan

(b)

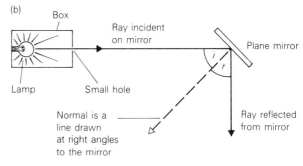

Box

Ray incident on mirror

Plane mirror

Lamp

Small hole

Normal is a line drawn at right angles to the mirror

Ray reflected from mirror

Fig. 14.1 Light travels in a straight line until it is reflected by the mirror

The two laws of reflection

Looking at Fig. 14.1(b) we find that:

(a) the reflected ray of light still travels in the same (horizontal) plane as the incident ray and the normal,

(b) the *angle of incidence* is exactly equal to the *angle of reflection*, in symbols:

$$i = r$$

The reflection discussed so far, which is the reflection from mirrors, is called *regular reflection* and the two laws above are the laws of regular reflection. *Diffuse reflection* is explained on page 159.

Images seen in a plane mirror

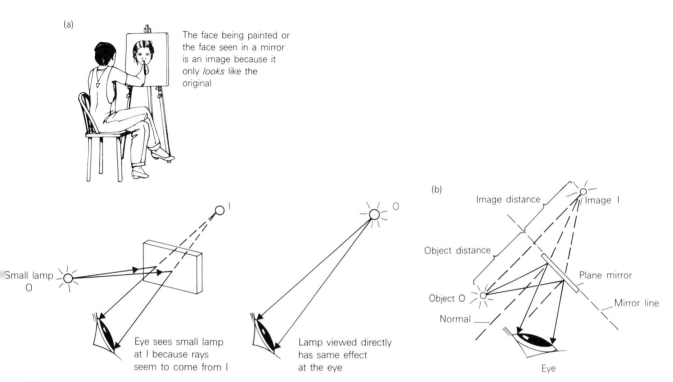

(a) The face being painted or the face seen in a mirror is an image because it only *looks* like the original

Small lamp O — Eye sees small lamp at I because rays seem to come from I

Lamp viewed directly has same effect at the eye

(b) Image distance — Image I — Object distance — Object O — Normal — Plane mirror — Mirror line — Eye

Fig. 14.2   Image formation: (a) the difference between an object and its image;   (b) image formation in the plane mirror

An *image* is something that *looks* exactly the same as an object but differs from it because only the object is tangible (Fig. 14.2(a)).

In Fig. 14.2 (b) rays of light enter the eye just as if they had been sent out from the place labelled I. This is where the eye sees the lamp when it looks towards the mirror. The image is therefore behind the mirror. Careful drawing of two rays from O to the mirror and marking in the reflected rays at the correct angles ($i = r$, remember) shows us that the position of the image is:

(a) as far behind the mirror as the object is in front (object distance equals image distance),

(b) the image I is exactly opposite to the object O, i.e. the line O to I is perpendicular to the mirror line.

For an object other than a small one the image is formed as in Fig. 14.3 (overleaf).

The image of each part of the object is as far behind the mirror as the part concerned is in front of the mirror.

The image has the following characteristics:

(a) it is the same size as the object, *and*

(b) is the same way up, as we see in Fig. 14.3(a), *but*

(c) the image is reversed from side to side (laterally inverted) so that the left of the object is seen as the right-hand side of the image. Thus the rose in Fig. 14.3(c) is seen near the right shoulder of the object, but near the left shoulder of the image. In Fig. 14.3(b) the girl will apply her lipstick with the right hand, but her image will appear to be left-handed.

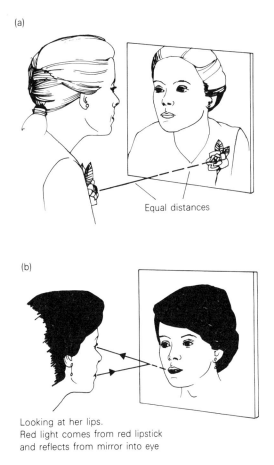

(a)

Equal distances

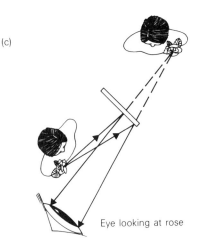

(c)

Eye looking at rose

(b)

Looking at her lips.
Red light comes from red lipstick
and reflects from mirror into eye

Fig. 14.3 Images in plane mirrors: (a) the position of the image;   (b) looking at oneself;   (c) object and image positions as seen from above

The looking-glass mirror

Mirrors can be made entirely from metal because metals are good reflectors of light; the necessary regular reflection is obtained by polishing the metal surface until it is very smooth. These mirrors suffer from tarnishing (discoloration due to chemical reaction with chemicals in the air), and so it is more common to use a thin metal coating applied to the very smooth surface of a sheet of glass. A mirror like this is cheap to make and, with a protective layer of paint over the metal coating, does not suffer from chemical attack and scratching is avoided.

Light passes through the glass, is reflected by the metal, and passes out through the glass sheet.

The size of mirror needed

The plane mirror needs to be only half the size of the image that is to be looked at. For example, a person 1.6 m tall needs a mirror 0.8 m high, placed about 0.8 m up from the floor in order to see her whole length from head to toe (Fig. 14.4 (a)).

The width of the mirror also needs to be only half of the person's width.

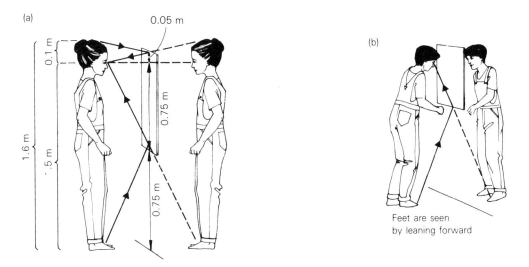

Fig. 14.4 Seeing one's full length in a mirror: (a) the minimum size for the mirror; (b) this mirror is too small

Using a mirror to create the impression of space

If a large part of a wall is glazed, an image of the room of equal size is seen in the mirror so that the room can appear twice its real size.

Inclined mirrors

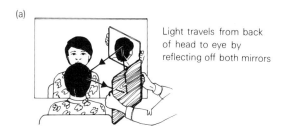

Light travels from back of head to eye by reflecting off both mirrors

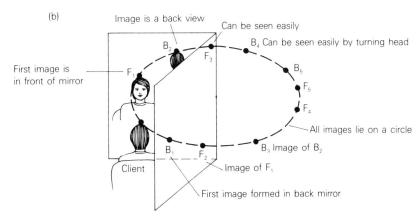

Fig. 14.5 Using two mirrors inclined to each other: (a) a small hand-held mirror close to the head; (b) two large mirrors

157

To show a client a view of (say) the back of her head, a second mirror is usually held, as shown in Fig. 14.5(a).

If two large mirrors were used in an attempt to provide the client with a rear view, both mirrors perhaps being permanently mounted, then the effect could be rather bewildering and not very successful. Fig. 14.5(b) shows that many images could be seen in the mirror and the client would probably see several with little or no movement of the head being needed. If the mirrors are almost parallel, a very good view of the back of the head could be obtained; but who would want to be faced with numerous people all identical to oneself?

Curved mirrors

These are of two kinds, *convex* and *concave*. A convex mirror has its shiny, reflecting surface on its inside, and a concave mirror is shiny on its outside (Fig. 14.6(a)).

A curved mirror intended for image formation is usually spherical in shape so that it has a centre of curvature as seen in Fig. 14.6(b).

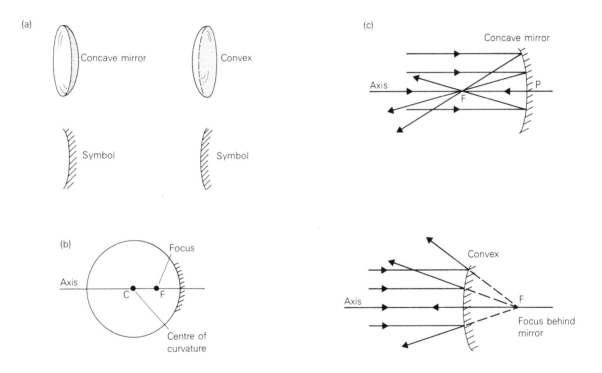

Fig. 14.6 Curved mirrors: (a) the two types of curved mirror; (b) curved mirrors are usually spherical; (c) the focus of a concave and convex mirror

The focus of a curved mirror

If rays of light arriving at a concave mirror are all parallel to each other and to the axis of the mirror, then they meet after reflection at a point called the *focus* (F) of the mirror (Fig. 14.6(c)). The distance FP between the focus and the mirror is called the *focal length* (abbreviation *f*). *f* is found to be equal to half of the mirror's radius of curvature.

In the case of a convex mirror, rays arriving parallel to the axis, after reflection do not pass through a point but do travel as if they had come from a point behind the mirror. This point is called the focus of the convex mirror.

A make-up mirror

A magnifying mirror for use in applying make-up or in shaving, is a concave mirror. This is held so that the face is a little nearer to the mirror than the focus and the image seen is upright and laterally inverted as in a plane mirror, but it is also *magnified.* Fig. 14.7 shows how rays appear to come from behind the mirror, creating the appearance of a bigger face.

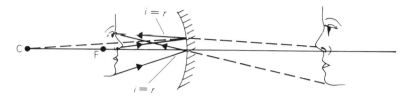

Fig. 14.7 How a make-up mirror works

A security mirror

For this purpose a large convex mirror is suitable. As seen in Fig. 14.8, a small, upright image is produced and we have the advantage that, with everything seen small, nearly the whole salon can be seen in the mirror. Furthermore, the mirror can be viewed from almost anywhere in the salon. It is useful not only for security purposes but for effective management.

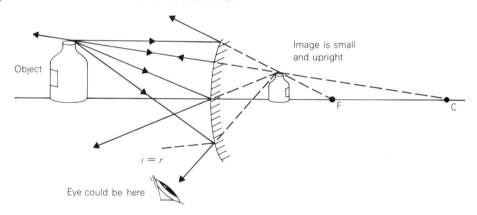

Fig. 14.8 Image formed by a convex mirror

Diffuse reflection

The reflection from a shiny metal surface is regular or mirror reflection, but what about unpolished surfaces such as writing paper, matt paint and cotton towels? Studied under a microscope these materials do not have smooth surfaces.

Light reflected from such materials comes away in all directions even though it may arrive in one direction only (Fig. 14.9 (a) overleaf).

It is usually desirable to choose make-up to be *diffusely reflecting,* certainly not too glossy. A high-gloss make-up acts like a mirror and a nearby light, such as the candle flame in Fig. 14.9 (b) (overleaf), would be imaged in the face.

A surface that is neither perfectly diffusing nor perfectly regular in its reflection gives a faint and hazy image of nearby lights, which we normally describe as a 'shine' or 'gloss'.

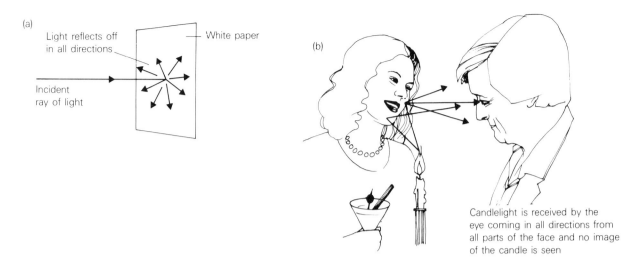

Fig. 14.9 Diffuse reflection: (a) from a sheet of white paper; (b) from matt-finish (non-glossy) make-up

REFRACTION AND LENSES

Refraction of light

So far we have found that the direction of travel of a ray of light is changed only by reflection, but a change of direction occurs also if light passes from the medium in which it is travelling, such as the air, into a different medium, such as glass. This phenomenon is called *refraction*.

As shown in Fig. 14.10(a), the ray is turned towards the normal if it passes from air into glass, and turns away from the normal if it passes from the denser medium into the less dense one.

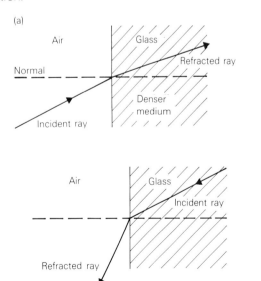

Fig. 14.10 Refraction of light: (a) how the light changes direction; (b) refractive index (n); (c) ray perpendicular to surface

Refractive index of a surface

The *refractive index* of a surface tells us how effective it is in producing refraction.

To be precise, the refractive index (n) is equal to x divided by y in Fig. 14.10(b) and equals 1½ for light going from air into ordinary glass.

If the incident light ray arrives perpendicular to the surface, it passes straight through.

($x = 0$ so that y ($= n$ times x) $= 0$ also.)

Refraction by a glass prism

The ray of light drawn in Fig. 14.11 is refracted as it enters the *glass prism* and then, as it passes out of the glass into the air again, it suffers a second refraction which adds to the deviation of the ray.

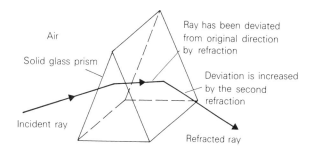

Fig. 14.11 Refraction by a prism

Refraction by lenses

Lenses, like curved mirrors, can be concave or convex in shape (Fig. 14.12(a)). A convex or converging lens brings rays closer together. A concave or diverging lens makes rays separate more.

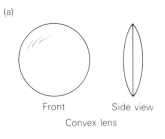

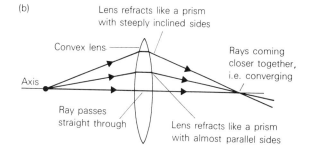

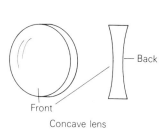

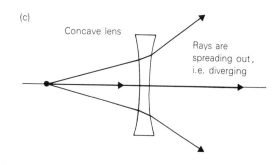

Fig. 14.12 Lenses: (a) types of lenses;   (b) refraction by lenses

Focus of a lens

The focus of a convex lens is the place where rays meet after refraction if the rays all arrive parallel to the lens axis (Fig. 14.13 (a)).

The focus of a concave lens is the point where rays seem to come from after refraction if the rays all arrive parallel to the axis (Fig. 14.13 (b)).

The distance from the lens to the focus (on either side of the lens) is the focal length.

A powerful lens is one that causes a lot of refraction and therefore has a short focal length. The lens' surfaces are very curved.

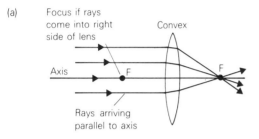

(a) Focus if rays come into right side of lens

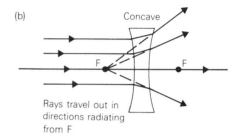

(b)

Fig. 14.13   The focus of a lens

Use of a convex lens for epilation

A large convex lens is used to magnify the hair and mouth of the hair follicle where a therapist is to insert the epilation needle. The lens is usually surrounded by a tubular lamp to provide adequate illumination and fixed on a stand so that it looks like the one shown in Fig. 14.14 (a).

The *epilation lens* should be placed so that the hair lies a little nearer to the lens than the focus. The exact position required is such that the image is about 25 cm (10 inches) from the eye because this is where the eye sees things in clearest detail.

In Fig. 14.14(b) two rays are shown travelling from the top of the hair through the lens and into the eye. These two rays have been chosen for the diagram because we know that the one arriving parallel to the axis must pass through F and the one arriving at the centre of the lens passes almost straight through. Extending the directions of the two rays away from the eye shows where the top of the image is. The whole image can then be marked in as shown. It is large and upright. The lens is acting as a magnifying glass.

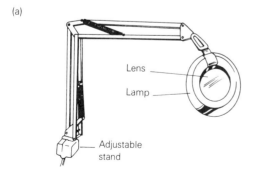

(a)

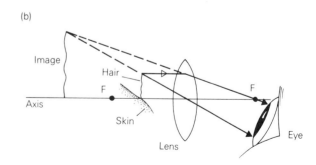

(b)

Fig. 14.14   A convex lens for epilation work: (a) appearance of lens on stand;   (b) the convex lens is acting as a magnifying glass

The epilation lens as a fire hazard

When a convex lens is not being used it should preferably be covered up, but otherwise it must *not* be left where the sun could shine through a window and reach the lens. Parallel rays from the sun would be made to converge to the focus of the lens. If a piece of paper or cloth found itself placed at this spot, it would become heated by so much energy arriving there. It could burst into flames (Fig. 14.15).

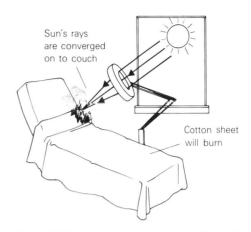

Fig. 14.15 A fire started by an epilation lens

The camera

It could be useful for a beauty therapist to know a little about *cameras* and, while we are discussing lenses, it seems worthwhile to explain the basic principles on which cameras work.

When the object is at quite a large distance from the camera an image is formed on the photographic film, as shown in Fig. 14.16. For a sharp image the distance from the lens to the film must equal the focal length. This means that all the white light arrives at one place on the film (the face of the image), all the red light in another place, and all the blue light at the legs of the image.

If the object is nearer to the camera (i.e. 'close up'), then the lens must be screwed forward to obtain a sharp image on the film.

The various parts of the film are affected by the amount of light received. This can be controlled by choosing a suitable time for which the film is exposed to light, i.e. the speed of the shutter must be chosen so that it opens for a suitable length of time. Also the aperture size can be selected (*f*/5.6, *f*/8, *f*/16, etc.) to let a greater or smaller amount of light in during the time of exposure.

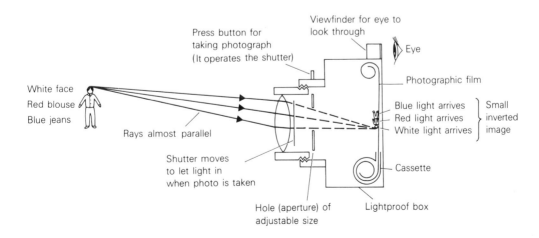

Fig. 14.16 The main features of a simple camera

Each picture taken uses an area of the film called a *frame*, and each frame is moved on after it has been exposed. When the whole film has been exposed, it is wound back into the lightproof cassette. It can then be removed from the camera and taken to be *processed*.

The processing of a black and white film involves washing it in a *developer* which turns the film black where light has been received. The film is then washed in *fixer* which prevents the film being affected by any more light. The film can then be taken into the light. The film is a *negative* because a white face is black on the film while dark clothing appears white. The other part of the processing operation is *printing*. For this, light is made to pass through the negative on to a sheet of photographic printing paper. Where the negative is dark, little light will pass through, and so the printing paper will be light in colour after developing and fixing it. Thus the print image is a positive image, the face being white as it was originally.

Real and virtual images

An image is 'real' if light actually arrives at the image as in the camera.

The other examples of images discussed in this chapter, namely the plane mirror, make-up mirror, security mirror and the magnifying glass, were ones in which the image is behind the mirror where no light reaches, or behind the hair and skin (Fig. 14.14(b)) where no light arrives. Such images are described as *virtual*.

How the eye works

When light from an object enters the eye it is refracted by the lens (and, in fact, this is helped by the cornea) and forms an image on the retina. This is illustrated in Fig. 14.17.

For simplicity the figure suggests that the lens only produces refraction although the cornea is also a convex lens.

The image formed on the retina is real (and inverted). The lens' focal length must be adjusted by the action of the eye lens muscles to obtain the correct amount of refraction to give a sharp image.

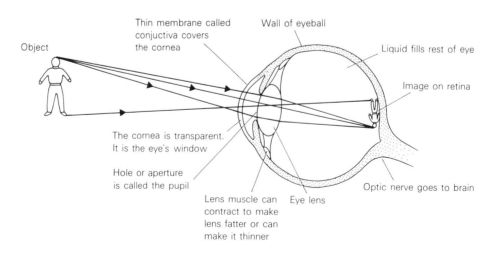

Fig. 14.17   Refraction by the eye

Spectacles

Long sight and *short sight* are defects of the eye caused by the eye's producing too much or too little convergence of the light rays. Spectacles can be used to overcome these problems. Spectacles with convex lenses are used to increase the convergence; concave lenses are used to decrease the convergence.

SUMMARY

- Light usually travels in straight lines, each part of the light emitted from a lamp travelling outwards as a ray.

- A mirror can change the direction of a ray. The laws of reflection tell us that the angle of incidence equals the angle of reflection.

- The image formed in a plane mirror lies as far behind the mirror as the object is in front. The image is virtual, laterally inverted (if the mirror is vertical), the same size as the object, and the same way up.

- The minimum size for the mirror if the client is to see the whole of her body is half her height and half her width.

- A second mirror enables a rear view or side view to be given to a client, but large mirrors may result in too many images.

- The focus of a concave mirror is the point through which all reflected rays pass if the rays came to the mirror parallel and close to the axis of the mirror.

- A concave mirror used as a make-up mirror is held so that the face is a little nearer to the mirror than its focus.

- A convex mirror can be used for security and effective management.

- When reflection is diffuse, light leaves the reflecting surface in all directions. No clear image is possible. Writing paper, matt paint and non-glossy face powder are diffuse reflectors.

- Refraction is the change of direction of light when it passes from one medium into another, for example, from air into glass. It is employed in lenses and prisms.

- A convex lens can be used as a magnifier for observing the hair and follicle for epilation.

- A convex lens as used for epilation must not be left near a window or it may converge the sunlight and set fire to nearby flammable materials.

- A camera uses a convex lens to produce a real, inverted, small image on a photographic film or plate. The image is later made visible by development and then made permanent by fixing.

- The eye works like a camera. Light enters the cornea, is focused by the lens, and so forms a real image on the retina.

- Convex and concave lenses are used in spectacles to overcome eye defects.

EXERCISE 14

1. State the two laws of reflection.

2. What is *diffuse reflection*? Give an example of a diffusely reflecting cosmetic.

3. Describe the position and size of the image formed in a plane mirror.

4. What are the size requirements of a plane mirror for a client to see the whole of her body?

5. What is the advantage of using two mirrors rather than just one? What problems arise if the mirrors are too large?

6. (a) What is a concave mirror? How is it used as a make-up mirror?

(b) Suggest a use in the clinic for a convex mirror.

7. How does the image in a convex mirror differ from that in a plane mirror? (CGLI)

8. Define the *focus* and *focal length* of a concave mirror.

9. What is refraction of light?

10. Define the *focus* of a concave lens.

11. For what purpose are convex lenses used in many beauty clinics?

12. Name a particular hazard associated with convex lenses.

13. Compare the image of a person in a camera with the image seen in a plane mirror on the wall.

14. What are the two stages of processing to obtain a print after a photographic film has been exposed?

15. What are the functions of: (a) the retina, (b) the lens, (c) the cornea of the eye?

15

THE ELECTROMAGNETIC SPECTRUM AND COLOUR

THE ELECTROMAGNETIC SPECTRUM

An experiment to demonstrate a spectrum

If a ray of light from an ordinary, incandescent light bulb has been refracted through a prism and is allowed to fall upon a white screen, the spot of light obtained is edged with red on one side and blue* on the opposite side (Fig. 15.1(a)).

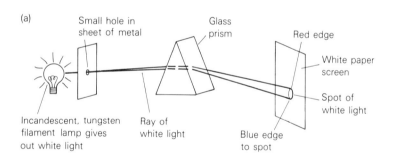

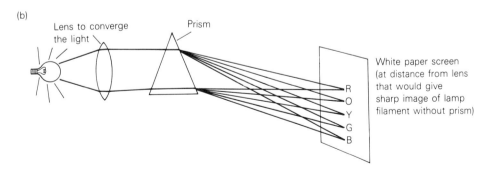

Fig. 15.1   Obtaining a spectrum: (a) the separation of colours from a ray of white light;   (b) obtaining a clear (fairly pure) spectrum

---

*Strictly speaking, violet.

The separation of colours from the ray of white light is most noticeable at large distances from the prism, but usually at such distances the ray has broadened so that the light is thinly spread. Fig. 15.1(b) shows a better method for displaying the colours and the white light from the lamp is seen to be split into a whole range of them. We describe these as red, orange, yellow, green and blue, in this order. We don't usually worry about giving names to other colours, except perhaps to divide blue into blue, indigo and violet.

White light is made up of red, orange, yellow, green and blue light.

The display of the colours of which the original light is composed is called a *spectrum*. A spectrum in colour is shown in the coloured plate (frontispiece).

A rainbow is a similar spectrum which again shows that white light, in this case from the sun, is made up of these colours.

Electromagnetic radiation

Because light is thrown out of a lamp, in fact at great speed, the word *radiation* is appropriate. But why electromagnetic? The answer is that light is made up of alternating electric and magnetic fields.

An electric field is what is present at a place where an electron experiences an attraction or repulsion. Usually the field is due to there being one or more charged objects nearby (Fig. 15.2(a)).

A magnetic field is what causes a place to affect a compass needle (Fig. 15.2(b)).

There is, of course, no sign of charged objects or of magnets in a ray of light and it is not appropriate to attempt explaining the cause of the alternating fields.

Since it is the alternating electric field in the light that enables us to see and which makes the camera work, the magnetic field is considered less important, and it has not been drawn in Fig. 15.2(c).

Looking at Fig. 15.2(c), it is easy to see why light is often called a *wave*. The electromagnetic wave, for that is what light is, does resemble water waves on the sea.

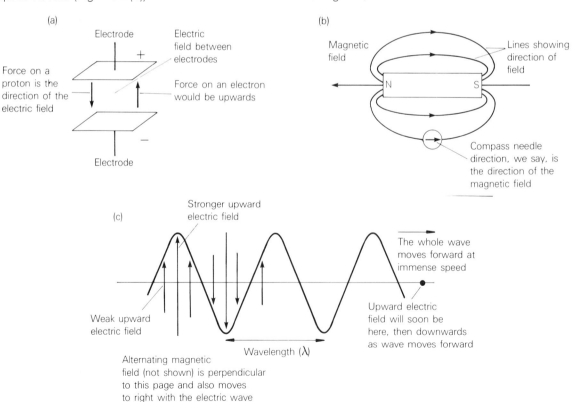

Fig. 15.2 Electromagnetic waves: (a) an electric field; (b) a magnetic field; (c) light consists of alternating electric and magnetic fields

The latter are characterised by a regular fluctuation of the height of water, a rise then a fall alternately, whereas a light wave has its electric field directed upwards and then downwards alternately as in the figure. Like waves on water the electromagnetic waves move forward away from their source, the direction of travel being perpendicular to the alternating electric and magnetic fields.*

Wavelengths of light

Referring again to Fig. 15.2(c), we see that the *wavelength* of light is the distance from one place to the next place along the direction of travel where exactly the same thing is happening, i.e. from where the electric field is strong and upward to the next place where the field is the same. Wavelength is usually denoted by the Greek letter λ (lambda).

The wavelength of light is always between 400 and 700 nanometers, a nanometer (nm) being a thousand-millionth of a metre. Its smallness is no doubt surprising.

Within this range of wavelengths we have all the colours of the spectrum with 400 nm being the blue end of the spectrum and 700 nm the red end.

Wavelengths can also be measured in *angstrom* (Å): 10 Å = 1 nm.

Light energy

When light passes through any medium, electrons in its path will experience alternating upwards and downwards forces from the alternating electric field. If these electrons are sufficiently free to move, they will therefore do so. Thus it is that electromagnetic radiation can give energy to materials that they enter. We realise then that electromagnetic radiation is itself energy.

Electromagnetic radiation is another form of energy.

Frequency of an electromagnetic wave

The *frequency* of an electromagnetic radiation, usually denoted by *f*, is the number of times the electric field alternates per second, i.e. the number of cycles per second or (as explained in Chapter 11) the number of hertz (Hz).

When light is arriving, for example, at the eye, the electric field reverses once in every wavelength that is received. This means that *f* wavelengths arrive per second so that

Velocity of radiation = Wavelength × Frequency

This is the same equation as was used for the speed of sound (p. 119).

Infra-red radiation (IR)

If a thermometer with a blackened bulb (as used in Fig. 10.12, page 110) is held just in front of the various colours on the screen in Fig. 15.1(b), it will be warmed by the radiation, but if it is moved beyond the red end of the spectrum the heating is surprisingly greater (see Fig. 15.3, overleaf). Clearly, an invisible radiation is arriving and its wavelength is greater than 700 nm. This radiation is called *infra-red radiation* and is what we have previously called radiated heat.

Infra-red radiation is electromagnetic radiation with a wavelength longer than that of red light. It is the same thing as invisible radiated heat.

---

*The truth is that the electric fields are not always up and down. The light from most lamps has its electric fields up and down at one instant, but after many alternations, which are amazingly rapid, the field may be alternating from side to side and, later again, in a direction tilted to the horizontal. However, this is not important to us.

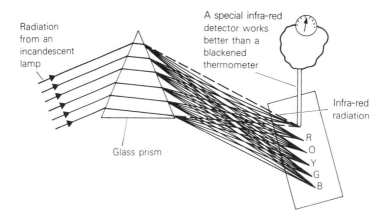

Fig. 15.3   Infra-red radiation

Ultra-violet radiation (UV)

Beyond the blue end of the spectrum another radiation is found. This is *ultra-violet radiation* (abbreviation UV). It lies in the spectrum beyond the violet or, as we have called it, the blue. Its presence can be shown by using a piece of photographic printing paper as the screen (see Fig. 15.1(b)). After development and fixing of the print, it will be seen that some radiation has arrived and blackened the print on an area that extends well beyond where the visible spectrum ended (Fig. 15.4).

An alternative method of detecting UV radiation is to arrange that it is changed into visible light when it arrives at the screen. Materials are available that glow green when struck by UV light. Many writing papers contain chemicals, called *phosphors,* that convert UV into blue light.

Ultra-violet radiation is found in the spectrum beyond the blue. Its wavelength is shorter than blue light.

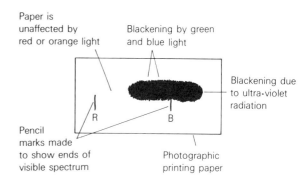

Fig. 15.4   Detecting ultra-violet radiation photographically

---

The complete electromagnetic spectrum

Not only visible light, infra-red and ultra-violet radiations are electromagnetic but also X-rays, gamma rays from radioactive materials, microwaves used for cooking and by physiotherapists for heat treatments, and radio waves.

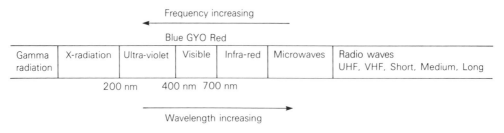

Fig. 15.5   The complete electromagnetic spectrum

Although these other radiations are of no importance in the beauty clinic, it should be realised that they differ from visible, IR and UV radiations only by having different wavelengths. As seen in Fig. 15.5, X-rays have shorter wavelengths (higher frequencies) than ultra-violet, while radio waves have large wavelengths and, in comparison with visible light, much lower frequencies. Most readers are familiar with the division of radio waves into long waves (lowest frequencies), medium waves, short waves, very high frequencies (very short wavelengths, compared to other radio waves) and ultra-high frequencies (UHF).

COLOUR

Studying colour

The beauty therapist and beautician should learn about *colour* because of its relevance to make-up. Colour is also of importance when choosing suitable decor for the clinic or salon and when using safety signs.

How the eye sees colours

The *retina* of the eye is made up of a large number of very small areas which are the ends of *rods* and *cones*. There are three kinds of cones. We will say nothing about the rods for the moment.

When an image is formed on the retina, red light arriving here, green light somewhere else, no light arriving at some other parts, the red light will affect cones of one type (called *red receptors*) wherever it arrives. Similarly, wherever green light arrives, the second type of cones (the *green receptors*) are affected, and any blue light will affect the third type of cones (the *blue receptors*).

The retina contains red, green and blue receptors.

Now these receptors are also affected, although to a smaller extent, by other colours of light than their own colour. For example, light with a wavelength near to red (say) orange, will affect the red receptors quite strongly but the green and blue receptors less.

The receptors then send messages (via the *optic nerve*) to the brain so that it can discover what colour of light, and how much of it, has arrived at the various parts of the retina.

If white light falls on a part of the retina, then all three types of cone in that area will be stimulated equally.

Making the eye see white

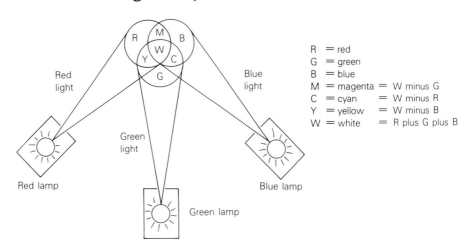

R = red
G = green
B = blue
M = magenta = W minus G
C = cyan = W minus R
Y = yellow = W minus B
W = white = R plus G plus B

Fig. 15.6 Mixing three primary colours of light (see also the frontispiece)

Proper white light, which means light containing all wavelengths, affects the red, green and blue receptors equally. Now if equal amounts of red, green and blue light are mixed together and reach a part of the retina, the brain will receive the same messages as it gets from the arrival of proper white light although the eye is not receiving all wavelengths (Fig. 15.6).

We need to mix three colours to make the eye see white.

The three colours used are called primary colours.

The primary colours of light are a red, a green and a blue colour.

Complementary colours of light

If cyan is mixed with red light we can get white light (white as seen by the eye) because one of these makes up for, or *complements*, what is missing in the other light. Cyan light is *complementary* to red light. Similarly magenta is complementary to green, and yellow is complementary to blue light. We can write all this as three equations:

$$\mathbf{C + R} \ (= \mathbf{W - R + R}) = \mathbf{W}$$
$$\mathbf{M + G} \ (= \mathbf{W - G + G}) = \mathbf{W}$$
$$\mathbf{Y + B} \ (= \mathbf{W - B + B}) = \mathbf{W}$$

A primary colour added to its complementary (or secondary colour) gives white light.

Filters

A *filter* is named according to the colour of the light it lets through. For example, a red filter lets only red wavelengths through and a yellow filter only yellow wavelengths.

The light that does not get through is *absorbed* (meaning 'soaked up'). The energy of this light has been entirely converted into heat in the filter.

To obtain a red light from a lamp producing white light, the light can be made to pass through a red filter, and this is useful for lighting effects on the stage, in the salon and for advertising displays.

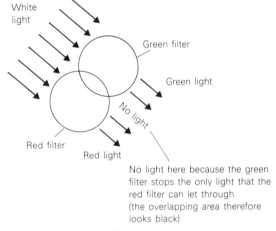

Fig. 15.7 Colour filters work by removing colours

Rod vision

When the eye receives very little light, usually because it is night time, the cones do not work and the eye sees by using the rods in the retina. Rods can only respond to the amount of light energy received and cannot distinguish colours. Thus we see in black and white only.

The colour of a surface

The colour of green grass, a yellow lemon or of a red dress is decided by the colour of light which comes from it when it is viewed under white light. Grass is green because it reflects the green light but absorbs the other colours arriving in the white light. So the eye sees green. The surface of a lemon absorbs blue light from the white light so that the lemon looks yellow.

A white surface reflects (diffusely) all light that falls upon it. A mirror does too but not, of course, diffusely.

Dyes and pigments

Dyes are coloured solutions, in water or some other solvent. A good example is *eosin* which has proved useful in lipsticks because it stains the lip surface red.

Pigments are insoluble, solid colouring materials that are used in paints and in cosmetics such as lipsticks, rouges, mascaras, etc. Examples are black carbon (the same

thing really as soot), naturally occurring iron oxide which can be yellow (ochre), red or brown, depending on the other chemicals present with it; another oxide of iron is black.*

Primary colours of pigments and other colorants

A red pigment reflects red light and some of the colour next to it in the spectrum, i.e. orange. A yellow pigment reflects yellow light and the orange and green next to it in the spectrum, and the blue pigment reflects blue and the green next to it.

> Red pigment reflects RED and ORANGE and absorbs YELLOW, GREEN and BLUE.
>
> Yellow pigment reflects ORANGE, YELLOW and GREEN and absorbs RED and BLUE.
>
> Blue pigment reflects GREEN and BLUE and absorbs RED, ORANGE and YELLOW.

Because pigments work by absorbing some colours we must emphasise the colour which a pigment removes in order to explain its working.

The three pigments mentioned above, red, yellow and blue, are the three primary pigments because suitable mixing of these three pigments can produce any colour (other than white).

Mixing pigments to obtain other colours

If blue paint is mixed with yellow paint, a green paint can be made because blue absorbs red and yellow. Yellow paint absorbs red and blue so that only green light is not absorbed.

In the same way, red pigment mixed with yellow pigment produces orange. Of course, this is only true if enough of the second colour is added.

Note that the primary colours of pigments (red, yellow and blue) are not the same as the primary colours of light (red, green and blue) and that the former add to produce black but the latter combine to produce white light.

The three primary pigments mix to give black.

The colour disc (Fig. 15.8) tells us the result of mixing any two pigments. The product is shown on the disc between the two colours that are mixed.

Purple is the colour (of a pigment) that reflects some red light and blue light.

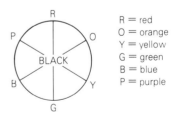

Fig. 15.8   The colour disc (see also the frontispiece)

---

*The term *pigment* properly used should apply to insoluble organic colorants. Lakes are insoluble, organic solids coloured by the use of a dye.

Complementary pigment colours

The result of mixing two colours that are opposite to each other on the colour disc is to produce black. They are *complementary pigments*. One absorbs what the other does not.

As an example, orange paint added to blue will make black paint, or, in practice, something close to it.

However, too much blue produces a blue or dark blue while too much orange gives an orange or brown (dark orange).

The influence of lighting colour upon surface colour

The colours of make-up, clothes or room decor can be very different when seen under different colours of light. Even small differences between lamps that are apparently white can affect the appearance of a coloured surface.

What colour will a blue eye-shadow appear when illuminated by the yellow (sodium lamp) lighting from a modern street lamp? Answer: black! The blue pigment of the eye-shadow absorbs all colours except blue and usually the green next to it in the spectrum, so that the yellow sodium light, which is a single wavelength, is absorbed. Other street lamps may be mercury vapour lamps (HPMV lamps) emitting a light which is deficient in red and this gives the skin an unhealthy bluish colour. Fluorescent lamps are low-pressure mercury vapour lamps with a fluorescent material (i.e. a phosphor) to convert the ultra-violet radiation produced into visible light. These are used for both indoor and outdoor lighting and their light is often deficient in one or more colours.

Ordinary tungsten filament lamps give a light which, though regarded as a white light, is definitely rather yellow in appearance, having a low proportion of blue in it. This light gives the skin a warm and pleasant appearance.

An example of the way lighting can affect the appearance of make-up can be seen in the coloured plate (frontispiece).

At this stage the reader may wonder what lighting shows the true (?) colour of a surface. The best answer is probably sunlight.

Pastel colours

If a coloured pigment is mixed with a white pigment we create a *pastel colour*.

Pink is a pastel colour obtained by mixing red with white. The red has been diluted with white.

A grey surface is one that reflects all colours equally but not very much. Black pigment mixed with white produces grey. A white surface receiving little light and therefore reflecting little may also appear grey.

Dark colours

These are obtained by mixing black with a colour.

Making allowance for the lighting colour

If everything in sight is illuminated by (say) pink light, the eye will make allowance for this in deciding what the colours are of the objects seen. We expect grass to be green, and if some other object at the same time looks the same colour as the grass, we would tend to say this object too is green. This helps us to see colours.

Choice of colours for room decor

Presumably it is because the sky is blue that we find blue walls or ceilings create the effect of spaciousness or of being outdoors. Also this colour gives the impression of coolness. In contrast, red reminds us of fire and makes the occupants feel a room is warm, sometimes equivalent to a temperature rise of several degrees. Green for the walls is relaxing and purple can seem regal and expensive.

Perhaps it is because of our familiarity with green grass that we do not object to having other colours adjacent to grass green. It is generally said that adjacent colours on the colour disc (Fig. 15.8) blend with each other, but colours opposite to each other on the colour disc (complementary colours) produce strong contrast when seen side by side.

Safety and colour

The agreed background colours for safety signs are red for warning against dangers, blue for signs giving orders that must be obeyed, and green indicating safety (for example, exits and directions to a first aid area).

For these signs to be fully effective their appearance must not be changed by their being seen under unusual colours of lighting.

Colour can also help to promote safety if obstacles are painted to make them stand out against their background colour. Yellow and black stripes are suitable for some places.

SUMMARY

- White light is made up of all colours (that is, red, orange, yellow, green and blue), and this order is the order of increasing frequency or decreasing wavelength.

- The speed of light equals the product of wavelength and frequency.

- Infra-red radiation has a larger wavelength and smaller frequency than visible light.

- Ultra-violet radiation has a smaller wavelength and larger frequency than visible light.

- The complete electromagnetic spectrum comprises; gamma, X-rays, ultra-violet, visible, infra-red, microwave and radio waves, and this order is the order of increasing wavelength.

- The limits of the visible spectrum are 400 nm at the blue end and 700 nm at the red end.

- The retina contains three types of cones which respond to the three primary colours of light, namely, red, green and blue. White light stimulates all three types of cone.

- If red, green and blue light, in suitable or equal (?) proportions, arrive together on the retina the eye

sees white, even though no orange or yellow wavelengths are arriving.

- Rods in the retina are useful at night but cannot distinguish colours.

- Surfaces have the colour of the light which they reflect when illuminated by white light.

- Red pigment reflects red and orange, yellow reflects yellow, green and blue, blue reflects blue and green (i.e. its own colour and the light next to it in the spectrum).

- Any colour, other than white, can be obtained by suitable mixing of red, yellow and blue pigments. These three pigment colours are the primary pigment colours.

- The appearance of clothes, make-up and room decor is affected by the colour of the lighting, for example, blue pigment cannot appear blue if illuminated by yellow light.

- Blue decor is cooling, red warming, green relaxing, and purple looks expensive.

1. White light is a mixture of colours. Name these colours in order of their wavelengths. (CGLI)

2. What is the frequency of yellow light of wavelength 600 nm, given that the speed of light in air is 3×10^8 m/s (i.e. 300 000 000 m/s)?

3. Where are infra-red and ultra-violet radiations found in the electromagnetic spectrum?

4. Name the approximate wavelength limits of the visible spectrum.

5. Name the primary colours of light.

6. What are the primary colours of pigments?

7. What primary pigments mix to give green?

8. What colours are seen when:
(a) green light illuminates a red lipstick?
(b) yellow sodium light falls on a blue dress?
(c) yellow sodium light on a green surface?

16

CLINIC OR SALON LIGHTING

REQUIREMENTS OF SALON LIGHTING

Illuminance

Above all else we expect to have enough light in a room to see what we are doing.

The amount of light energy falling (per second per square metre) on what we are looking at decides the *illumination* or, a better term, the *illuminance*. The unit employed for illuminance is the *lux* (rather than joule per second per square metre).

For a clinic or salon an illuminance of about 400 lux should be adequate while for corridors and cupboards 200 lux is acceptable.

For fine work, such as inserting an epilation needle, a higher illumination is needed but this is usually provided by a lamp close to the epilation lens (Fig. 14.14, p. 162).

To achieve the required general illumination there must be enough lamps of suitable brightness and size and suitably placed. Light-coloured walls (a pastel colour) will add reflected light to the direct light arriving at the place where work is being done.

Insufficient light will not damage the eyesight but it can cause headaches and sore eyes. Also it can lead to poor workmanship and to accidents.

The colour of the light emitted is important too and this was explained in Chapter 15.

Glare

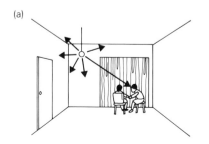
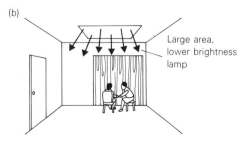

(a)

(b)

Large area, lower brightness lamp

Fig. 16.1 Discomfort glare: (a) the problem; (b) an answer

If, as one looks across the room, one object or part of the room is far brighter than the rest then we have the problem of *glare*. A simple example is an ordinary (incandescent) light bulb with no shade or luminaire fitted to it. One's eye is repeatedly attracted to this bright area and the effect is annoying. It is called *discomfort glare*, and can, like inadequate illumination, lead to poor work and indirectly to accidents.

The problem is greatest when the offending area is close to the direction in which you are looking. It can be reduced by using a lamp shade or a diffusing luminaire around the lamp, by changing the position of the lamp, or by using a lamp of larger area and lower brightness such as a fluorescent tube lamp. A brighter background also helps.

Shadows

Shadows in a room may be distracting and they may reduce the illumination where one is working. The use of several lamps can reduce shadows but large area lamps such as that shown in Fig. 16.1(b) are preferable.

On the other hand, a little shadow here and there adds interest to a room. Also some shadows may show the presence of a wall, pillar or obstacle that otherwise might not be noticed.

LAMPS AND LUMINAIRES

The incandescent, tungsten filament lamp

This is the ordinary light bulb, containing an electrically heated tungsten wire (the filament) inside the glass bulb. The filament glows because of its high temperature, i.e. is *incandescent*. The bulb is evacuated or contains a gas that will not allow the tungsten to burn away. The glass bulb may be clear or translucent (diffusely transmitting). See Fig. 16.2.

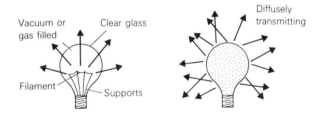

Fig. 16.2   A tungsten filament lamp

The fluorescent lamp

A typical fluorescent lamp fitting is shown in Fig. 16.3(a).

The principle on which this lamp works is that a current passes through the low-pressure mercury vapour in the tube (see Chapter 13) and produces mostly ultra-violet light. This ultra-violet radiation falls upon the special coating (the phosphor) on the inside of the glass tube and the phosphor absorbs the ultra-violet and in its place it sends out visible light, i.e. the phosphor *fluoresces*.

The phosphor is chosen so that its fluorescence is near to white light. Some lamps are deficient, for example, in red and others have had the red emission improved so that a warm (pinkish) white is produced.

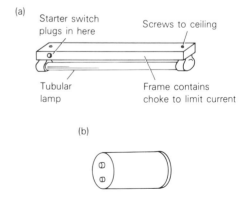

Fig. 16.3   (a) A fluorescent lamp and (b) its starter switch

When the lamp is first switched on it is necessary to warm it up for a few seconds by making current flow, not along the length of the tube, but through the electrodes (see Chapter 13). This current also passes through the starter switch (which is seen to glow) and heats a bimetal strip inside it. As a result of the strip's bending this switch opens, stopping the current through the electrodes. Only then does current pass along the tube and the lamp become fully operational.

Lamp shades and luminaires

The purpose of a *luminaire* is to prevent light travelling in unwanted directions and to send it instead to where it is most needed. A *lamp shade* is opaque and its inner surface reflecting. It is commonly used for a desk-top reading lamp.

Diffusing luminaires are most useful for reducing glare, at the same time improving the appearance of a lighting system.

Choice of lamp

The tungsten filament light bulb has the advantage of small size and with a suitable shade its light can be directed to a particular area. Glare can be a problem if the bulb has clear glass, although a glass or similar luminaire can be fitted around it. Such a lamp has a limited life because the tungsten slowly evaporates (and condenses on the glass surface) so that the filament thins and finally breaks. The colour of the light is usually acceptable.

Incandescent lamps can be obtained as striplights, i.e. as tubes rather than bulbs.

The fluorescent tube lamp is always long (longer for higher wattages), but it gives more light per unit of electricity used (a 40 W fluorescent giving out as much light as a 100 W bulb). The life of the lamps is long, but the starter switch can soon wear out if the lamp is switched on and off too often. However, the starter switch is easily unplugged and replaced. One disadvantage of the fluorescent lamp is that it takes a few seconds to get started. Where glare is concerned it is the obvious choice, especially if a luminaire is also used, as in Fig. 16.4.

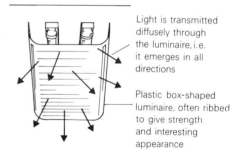

Light is transmitted diffusely through the luminaire, i.e. it emerges in all directions

Plastic box-shaped luminaire, often ribbed to give strength and interesting appearance

Fig. 16.4   A large-area luminaire

SUMMARY

- Illumination describes the amount of light falling on a surface. It is measured in a unit called the *lux*.

- Glare may prevent seeing (disability glare) or cause annoyance, distraction or eye tiredness (discomfort glare).

- Discomfort glare is due to a very bright area being seen against a less bright background.

- Discomfort glare from a small lamp may be avoided by use of a lamp shade, a diffusing luminaire, or a change in its position.

- The wavelengths, and so the colour produced from a fluorescent lamp, depends upon the phosphor used.

- Fluorescent lamps are cheaper to run (for the same amount of light) than incandescent lamps. However, they are bigger and take a few seconds to get started. A suitable illumination for a desk or couch in a clinic would be about 400 lux.

- A starter switch is needed to start low-pressure mercury vapour lamps such as the ordinary fluorescent tube lamp. They are easily replaced when they wear out.

EXERCISE 16

1. Suggest a suitable illumination for a clinic.

2. Name a treatment that will require especially good illumination.

3. Glare can cause strain and can frequently be a hazard in a salon. State *three* ways in which glare can be reduced or eliminated. (CGLI)

4. Why are some lamps described as *fluorescent*?

5. Name two advantages of a fluorescent lamp compared with an incandescent lamp.

6. What is the function of a starter switch?

17

INFRA-RED AND ULTRA-VIOLET RADIATIONS

INFRA-RED RADIATION

What is infra-red radiation?

As explained in Chapter 15, *infra-red radiation* is electromagnetic radiation which is invisible and has a wavelength longer than that of red light (i.e. longer than 700 nm), but shorter than the wavelengths of micro-waves and radio waves. It is precisely the same radiation as the radiated heat described in Chapter 10.

For many purposes it is appropriate to divide infra-red radiation into near and far IR, meaning the IR near to the visible part of the electromagnetic spectrum and the IR that is further from the visible. See Fig. 17.1.

Fig. 17.1 Near and far infra-red

Radiation from a hot object

A *hot object* radiates heat, i.e. emits infra-red radiation; if it is hot enough, it glows red-hot, or even yellow-hot, or (hotter still) white-hot. Remembering the experiment where a tungsten filament lamp was used and a spectrum produced, we realise that not only infra-red radiation and visible radiation was given out, but even ultra-violet.

The amount of radiation emitted at each wavelength by a heated tungsten filament is shown by the graphs in Fig. 17.2 (overleaf).

We see from these graphs that the white-hot wire gives out most energy and that less is emitted as the temperature is reduced.

It is also noticeable that an object that is only black-hot, say 100°C, is (according to the graph), giving out no visible light as one would expect. The red-hot object is giving out some red light, but little or no green or blue.

This explains its red appearance, but look too at the far greater amount of near IR radiation that this red-hot wire is releasing. The white-hot object is emitting all the visible wavelengths, including even enough blue for the wire to appear a reasonable white colour; it also emits a lot of near IR and some ultra-violet.

> A black-hot surface radiates far IR.
>
> A red-hot surface radiates near IR and red.
>
> A white-hot surface radiates near IR and all colours and ultra-violet.
>
> Too much IR on the skin causes burns.

It is because hot objects, (like an electric fire element), when they send out energy, emit mostly IR that we tend to regard only IR as radiated heat. The heating by the visible light emitted is small.

In beauty therapy the term 'radiant heat' means near infra-red mixed with red light.

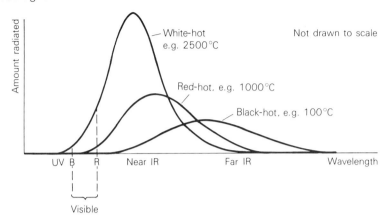

Fig. 17.2 The radiation from a heated tungsten filament

Some physical properties of infra-red

In many ways IR is like light. It can be reflected by polished metal surfaces and can be refracted, as in the prism experiment (Fig. 15.3, p. 170).

Near IR will pass through ordinary glass, but not far IR. (The IR detected in the spectrum produced by the glass prism was near IR.) A sheet of glass will stop the far IR reaching the heat detector in Fig. 10.13 (p. 111) but has little effect on the mostly near IR in the experiments of Figs. 10.11 and 10.12 (p. 110).

Physiological properties of infra-red

Near IR can penetrate the skin and reach a depth of some 5 mm, but far IR is stopped at the skin surface. The main effect of IR is heating, causing immediate erythema. The warming of the body is also very pleasant and relaxing. IR can encourage perspiration too.

Therapy using IR is aimed at producing some erythema. This is accompanied by increased blood and lymph flow and general increase in activity within the body tissue treated. Near IR is most effective but cannot be tolerated for long. The longer wavelengths of far IR are less effective but less irritant, and treatment can last for a longer time. The heating of the tissues by far IR depends, of course, on the heat produced at the skin surface spreading down into the deeper parts of the skin and the dermis by conduction and by convection (by blood flow).

Cataracts

These are opacities in the eye lens and it is known that IR radiation received in large and frequent doses can cause such eye defects. If one refrains from frequent staring into radiant heat lamps, then there should be no fear of harm to the eyes.

Heat also causes discomfort to the eyes because of its drying effect.

Sources of IR radiation

Two kinds of lamp are common. These are usually known as *radiant heat lamps* and as (true) *IR lamps*. The first of these radiates near IR and red, while the other lamp radiates far IR.

A radiant heat lamp

A typical lamp of this type is shown in Fig. 17.3.

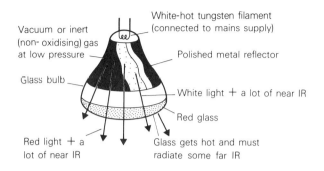

Fig. 17.3 A radiant heat lamp

In this lamp the electrically heated tungsten filament produces a lot of near IR radiation, and some white light (the small amount of ultra-violet is unimportant). The red glass stops other colours of light and the red, together with the large amount of near IR, passes out and is applied to the client at a typical distance of 50 cm.

The lamp wattage is usually about 200 W, and it requires only a few seconds to obtain full output of radiation.

Because the glass of the lamp becomes so hot, it should not be touched for fear of causing a burn. Another hazard is that, if the bulb is knocked or splashed with water, it may break and implode (like an explosion, but inwards).

The radiation from this lamp is largely near IR, and so it produces the quite deep heating characteristic of such wavelengths and the treatment should be kept short.

A (true) infra-red lamp

The use of the term 'infra-red lamp' is common practice when the lamp produces no visible light. However, since we have seen that the radiant heater emits mostly IR also, it seems to the author that the description 'true infra-red', to imply IR only, would be preferable. This non-luminous type of IR lamp may have the appearance shown in Fig. 17.4 and it emits long wavelength (far IR) radiation.

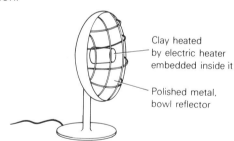

Fig. 17.4 A non-luminous IR lamp

The clay is very hot, but not so hot as to give out light. The comparatively small amount of radiation from a surface at such a temperature is made up for by there being quite a large area of clay surface.

The main hazard from this lamp is being burnt by touching it, especially because it looks the same whether it is on or off.

Because the radiation is long wavelength IR, it is less irritant than the radiant IR.

The curved mirror serves to reflect radiation that would otherwise not reach the client. Dents in this reflector can focus too much radiation to some areas of the client's skin and so must be avoided.

A few minutes are required for this kind of lamp to warm up.

The inverse square law

Light, or IR radiation, travels from a small lamp in straight lines which spread out (or *diverge*). Consequently, at a greater distance from the lamp the radiation is more thinly spread. The rule that describes this spreading is the *Inverse Square Law*. As seen in Fig. 17.5 (overleaf), doubling the distance from the lamp spreads the radiation out over a four times bigger area.

In an IR treatment, the dose received will be four times smaller if the distance between the lamp and the client's skin is doubled.

In fact, the rule applies *exactly* only to a lamp which is smaller than the lamp-to-skin distance, but it is

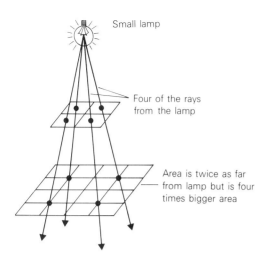

Small lamp

Four of the rays from the lamp

Area is twice as far from lamp but is four times bigger area

Fig. 17.5   The inverse square law

nevertheless a very good guide to the expected dose (or dose rate) from IR from a bulb-type lamp.

We can state the inverse square law as:

The dose doubles if $\dfrac{1}{\text{distance squared}}$ doubles

To illustrate the use of this law, suppose that the distance was 20 cm and we want to double the dose:

$\dfrac{1}{\text{distance squared}}$ is $\dfrac{1}{20^2}$ or $\dfrac{1}{400}$ and this must be doubled.

Therefore $\dfrac{1}{\text{new distance squared}}$ is $\dfrac{1}{200}$ which is $\dfrac{1}{14^2}$

approximately. So the new distance required is about 14 cm.

Thus the law tells us:

Doubling the distance reduces the dose by four times.

To halve the dose the distance must be multiplied by 1½ approximately.

If the distance is doubled but the same heating effect is required, the treatment time can be doubled.

ULTRA-VIOLET AND ITS USE FOR SUN-TANNING

Ultra-violet radiation

This radiation interests the beauty therapist because of its sun-tanning properties.

As already explained, UV is electromagnetic radiation that is invisible and has a wavelength between that of the blue (or 'violet') end of the visible spectrum and that of X-radiation.

Just as IR is divided into near and far according to the wavelength, so we divide UV into UVA, UVB and UVC. Of these, UVA is near-UV, i.e. it lies next to the visible part of the spectrum, UVB is next in order and UVC is the shortest wavelength UV (Fig. 17.6).*

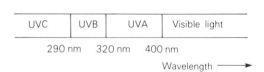

UVC	UVB	UVA	Visible light

290 nm   320 nm   400 nm

Wavelength ⟶

Fig. 17.6   The three divisions of ultra-violet radiation

Physical properties of UV

Because UV is similar to light, it can be reflected; it must be emphasised that reflection of UV can occur not only from mirror surfaces but also from other light-coloured surfaces.

UV can also be refracted, and this has been illustrated by the prism experiment (Fig. 15.1, p. 167 and Fig. 15.4, p 170).

---

*There is also vacuum UV lying between UVC and the X-ray part of the spectrum but this vacuum UV is unable to pass through anything but vacuum (not even the air), so that for this reason alone it would be of no concern to the beauty therapist.

The term cosmetic UV is sometimes applied to both UVA and UVB because they can each be used for cosmetic purposes (sun-tanning to improve the skin's appearance). UVC is not cosmetic.

As regards transmission through materials, UVA will, like visible light and near-IR, pass through ordinary glass, through Perspex (the best known acrylic plastic), and through fused silica (often called quartz glass). UVB will not pass through ordinary glass; some will pass through Perspex, but it is transmitted well by fused silica. UVC will not pass through ordinary glass or Perspex but most of it does get through fused silica.

Unlike visible light, UV will, if its wavelength is short enough, cause many chemical reactions to occur. The important example is the production of ozone from oxygen. The smell of small amounts of this gas can be noticed near to the high-frequency sparking apparatus and near to some UV lamps.

The wavelength of UV must be shorter than 250 nm (UVC) for ozone to be produced. The chemical change is brought about when UV splits oxygen molecules into separate atoms and these regroup to form some ozone molecules.

Physiological properties of UV

UVA can penetrate the *epidermis* and even the deeper layer of the skin, the *dermis*. UVB can penetrate the epidermis, but UVC does not even get far into the epidermis.

All can produce erythema and, with a higher dose, peeling or, finally, burning.

UVA and UVB are both able to give a sun-tan.

Other properties of UV include beneficial effects such as helping the body to produce vitamin D and killing germs; the possible undesirable effects are inflammation of the cornea of the eye (*keratitis*), ageing of skin and skin cancer. It is not yet known whether excessive UV can cause cataracts.

All of these properties are considered in more detail below.

Erythema

A mild *erythema* is very desirable to exercise the blood vessels and generally liven up the skin. UV like IR can provide this. A more intense erythema will be followed some time later by fine peeling of the skin (*desquamation*), due to the more rapid growth of the skin after the exposure to UV. This too can be beneficial. However, too much UV arriving on the skin in a short time causes burning with great soreness and blistering; this 'sun-burn' is definitely to be avoided.

Erythema is produced by all UV wavelengths but not all are equally effective in this respect.

The graph in Fig. 17.7 compares the ease with which erythema occurs with different wavelengths of radiation.

> *It takes about 1000 times more UVA than UVB or UVC to produce erythema.*

The reddening of the skin does not show up straight away when the skin is exposed to UV, although if erythema is obtained with UVA it can appear quite soon. With UVB it takes about an hour to show up and is most noticeable after about 10 hours, then fading, taking three

or four days to go completely. The UVB erythema is therefore delayed erythema. With UVC the erythema is similar to that of UVB.

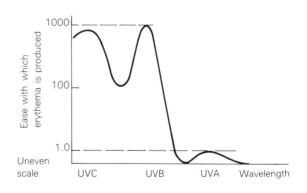

Fig. 17.7 Erythema caused by various wavelengths

185

Thickening of the skin

Another consequence of exposing the skin to UV is that the rate of growth of the skin is increased so that the skin becomes thickened; this increase may persist for some weeks after the exposure. This does help to protect the deeper parts of the skin from further doses of UV radiation.

The sun's radiation

The *sun* produces a spectrum like that of a white hot tungsten wire (Fig. 17.2) but, because of the sun's much higher temperature, it emits more UVA, UVB and even UVC. However, the UVC is all lost in the upper layers of the earth's atmosphere and so is absent from sunlight at ground level.

Sunlight consists of IR, visible light, UVA and UVB.

How effective the sun is for producing erythema, sun-tan, etc. depends upon atmospheric conditions and also upon the angle at which it reaches the skin (Fig. 17.8).

UVA and UVB are often referred to as *actinic* because they are found in sunlight whereas UVC is not.

Thick cloud in the sky stops UV but, what is not always realised, water droplets in the air in the form of thin cloud or mist may stop the sun's heat (IR) but do not necessarily stop UV which merely reflects off the droplets. What is more, this effect results in UV reaching the body from all directions so that a sun-shade does not stop it. Thus, you can become reddened, brown or burnt without even seeing the sun.

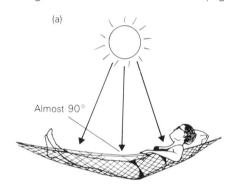 

Fig. 17.8   Sunbathing: (a) the maximum dose being received;   (b) very little sunlight striking the body

Sun-tanning

The purpose of *sun-tanning* is to increase the amount of the brown *melanin* pigment in the skin.

There are two kinds of sun-tan, as with erythema. It can either be *immediate* or *delayed*. Immediate tanning (IT) is produced only by UVA and, just like UVA erythema, requires high doses of UVA. The tan will appear within minutes but fades after a few hours or days. It seems to produce melanin from *melanosomes** that have been left in the skin as a result of previous sun tanning, or by a similar process. Delayed tanning actually causes fresh melanosomes to be produced in the skin, and the brown melanin is formed.

The effectiveness of UVA and UVB for delayed tanning (DT) is the same as for delayed erythema, occurring easily with UVB but requiring much larger doses of UVA.

About 1000 times more UVA than UVB is needed to produce sun-tan.

Tanning of the skin by sunlight is mostly due to the UVB radiation because UVB, although present in a smaller amount than UVA, is so much more effective in causing DT than UVA. The UVA will, typically, cause about one-sixth of the total tanning effect.

*See Chapter 20.

DT requires about ten hours to show up, usually giving the darkest tan after a few days, and it then fades over a month or so. After a few months a little is left as a permanent tan.

UVC produces very little tan, if any. In fact, it does us only harm and so is often described as *abiotic* (harmful to living things).

Finally, we all know that some people tan more easily than others. Fair-haired people are mostly fair-skinned and do not tan easily but can burn. A black-skinned person, whose skin is probably thick, has little need for sun-tan and is well protected from the sun, but large doses of UV can produce erythema.

Once some tan has been acquired this protects against erythema and sun-burn as well as other effects of UV.

Keratitis

Keratitis is inflammation of the cornea of the eye, and *conjunctivitis* is inflammation of the thin film that covers the outside of the cornea. These conditions are caused by UVB and by UVC, because these are stopped by the conjunctiva and cornea whereas most UVA passes through into the eye.

Keratitis is accompanied by pain, a feeling like 'grit under the eyelids', a dislike for light, and weeping of fluid around the eye. Fortunately these effects disappear within a few days with no apparent after-effects.

To avoid injury, eye protection is essential when UVB or UVC are being used. Even some clear spectacles will stop these radiations (glass, for example), but these radiations are often accompanied by intense visible light so that opaque eye covers are used.

It is felt by many people that eye protection should also be worn when UVA is being used, the justification being that, although UVA does not produce keratitis, there is no certainty that UVA is harmless to the eye. Some sun-glasses will stop UVA, but special spectacles can be obtained.

Ageing of the skin

It is to be expected that, as a person gets older, the skin becomes less elastic and shows wrinkles and folds; but these signs of *ageing* appear perhaps many years earlier if the skin has received large amounts of UVA. Sun-tanning should be undertaken in moderation.

Skin cancer

Repeated excessive doses of UVB are known to produce *skin cancer*.

Cataracts

Parts of the eye lens may become opaque (and so vision will be impaired) as a result of UVA radiation affecting the lens. This does not seem to be a common occurrence, which implies that the doses required to produce this condition are large and that photosensitisation is a prerequisite (see p. 188).

Vitamin D formation

UVB enables the body to produce *vitamin D* from chemicals normally present in the skin and close to the skin. Vitamin D is essential to the body and its produc- tion by UV is important, although it is obtainable from various foods (see Chapter 19).

Photosensitisation

The physiological effects of UV (including sun-burn) are often made much more noticeable by the presence of chemicals, including some cosmetics, on the skin, by drugs that have been taken, or even by eating certain foods (strawberries being an example often quoted). The phenomenon is called *photosensitisation*.

Some sun-tan preparations make use of photosensitisation because they contain bergamol oil, with the aim of making the sun-bather go brown more quickly. It is claimed by some people, however, that a very uneven tan can result and the unevenness can persist for years.

Sensitisation of the skin to UVB occurs if it is previously exposed to UVA. Strangely enough, however, if the UVA is applied after the UVB or UVC it reduces erythema produced by these shorter wavelengths.

Sun-tan ultra-violet sources

These are mostly of two kinds, namely, low-pressure mercury vapour (LPMV) tubes designed to produce UVA only, and high-pressure mercury vapour (HPMV) lamps that emit UVA, UVB and UVC. The UVA tubes are very common in sun-beds.

Sun-beds

These sun-beds, (looking something like the one in Fig. 17.9), produce sun-tan by use of large doses of UVA only or, perhaps with a little UVB included, with the intention that this should create melanosomes; thus the tan will not be limited by there being not enough melanosomes present.

Some tan is observable within minutes and a typical treatment session is half an hour, repeated every two days for a week or two. At the end of this time, some delayed tan can also be expected. Once again it is essential that the person has a skin that will tan in sunlight if this artificial sun-bathing is to work.

Erythema is possible as is sun-burn, but fortunately, many people get the immediate tan and finish the treatment session before the erythema shows up. Of course, the IT tends, anyway, to hide any slight erythema.

The lamps used are LPMV tubes. These produce almost entirely UVC as the result of electric current passing through the mercury vapour. The UVC then falls upon a layer of special phosphor coated on the inside of the glass tube. This converts the UVC into UVA which passes out through the glass tube and acrylic sheet to reach the sun-bather. R-UVA tubes have built-in reflecting surfaces.

UVA does not cause keratitis so that eye protection is not mandatory but, in case of long term effects on the eyes, suitable spectacles are worthwhile.

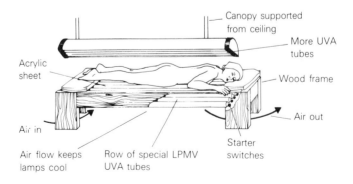

Fig. 17.9 A sun-bed with a canopy

HPMV sun-lamps

Two designs of *sun-lamp* are shown in Fig. 17.10.

The HPMV tubes shown give out a very intense light which is too much to look at. One would, therefore, want to use eye protection; indeed, this is essential because of the UVB and UVC radiations emitted.

The fact that UVA and UVB are produced in addition to UVC is the result of the high pressure of the mercury vapour. The higher the pressure, the more wavelengths we get, and at the pressures used a great deal of visible light is emitted. The tube becomes extremely hot and for this reason quartz is appropriate, rather than ordinary glass which would melt. Also the quartz allows the UVB (and UVC) to pass through . As with other lamps where a current is flowing through a gas or vapour, the current must be limited by a resistor or choke. In Fig. 17.10(a) it can be seen that a tungsten filament is used.

Sun-tan will largely be produced by the UVB radiation and the treatment time will be short.

As with the infra-red bulb, the inverse square law applies with reasonable accuracy because the lamps are neither long nor too big.

Note that, if the therapist does not wear eye protection, (thinking that she will not look at the UV lamp directly), she should realise that where light is reflected, from walls and light-coloured furnishings and clothes, UV is likely to be reflected also.

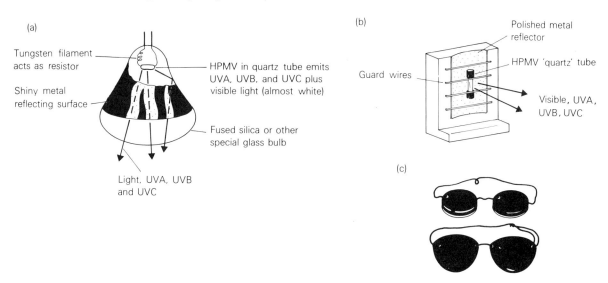

Fig. 17.10 Sun-lamps using HPMV lamps: (a) a bulb type; (b) an HPMV 'quartz' tube 'burner' with reflector; (c) goggles or eye pads *must* be worn

The UV steriliser

A *UV steriliser* is described in Chapter 22. An LPMV tube, with no phosphor needed, produces UVC radiation which is ideal for killing germs. The tube must be made of a material like fused silica that transmits the UVC radiation.

Ozone steamers

Because *ozone* has bactericidal properties (i.e. it can kill germs), it is common for it to be added to steam for skin treatments. The ozone can be created by using a high-pressure mercury vapour lamp. This quite small lamp emits ultra-violet radiation that converts some of the oxygen in the surrounding air into ozone. Air containing this ozone is drawn into the nearby flow of steam. See Fig. 17.11 (overleaf).

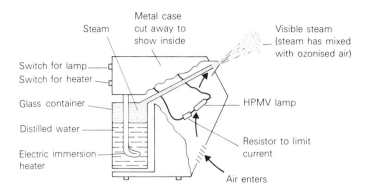

Fig. 17.11 The production of ozone in an ozone steamer

<hr />

SUMMARY

- Infra-red (IR) radiation may be divided into near-IR and far IR (which has a longer wavelength).

- A white-hot surface emits a lot of electromagnetic radiation, especially near IR and contains all visible colours.

- A red-hot surface radiates less. The radiation consists of IR plus some red light.

- A black-hot surface radiates still less and its radiation is IR (long wavelength) only.

- IR is invisible electromagnetic radiation and is the same thing as invisible, radiated heat.

- Near IR can pass through ordinary glass but not far IR.

- Near IR penetrates up to 5 mm of skin and underlying tissue and causes beneficial erythema.

- Far IR is less penetrating but is less irritant to the skin.

- IR improperly used can produce burns and, over a long period, cataracts in the eyes.

- A radiant IR lamp gives out red light as well as near IR.

- A (true) infra-red lamp gives out no light, only far IR.

- The inverse square law tells us that doubling the distance between lamp and client reduces the dose rate to a quarter of its previous value and so allows a four times longer exposure.

- UV radiation can be divided into UVA, UVB and UVC. Near UV is UVA.

- UVA and UVB can each be used for sun-tanning.

- UVA gives immediate sun-tan and is commonly used in sun-beds.

- UVB gives delayed tan which lasts longer.

- UVB can cause keratitis.

- Goggles must be worn when using UVB (and UVC).

- UV improperly used can cause sun-burn, keratitis and, over a long period, cataracts, ageing of the skin, and skin cancer.

- UVB can enable the body to produce vitamin D.

- UV, especially UVC, is germicidal.

<hr />

EXERCISE 17

1. Name the two regions of the invisible spectrum which are made use of in beauty therapy treatments. (CGLI)

2. Compare the wavelengths of radiation emitted by: (a) a white-hot surface, (b) a red-hot surface, (c) a black-hot surface.

3. Distinguish between a radiant lamp and a true infra-red lamp.

4. What advantage and what disadvantage has near IR compared with far IR for therapy?

5. Compare the sun-tanning properties of UVA and UVB.

6. What UV radiations are present in sunlight?

7. What are the hazards of UVB radiation?

8. Name a beneficial effect of UV other than sun-tan.

9. What application has the *inverse square law for light* in beauty therapy treatments? (CGLI)

10. Which beauty therapy treatment could result in conjunctivitis if certain precautions were omitted? (CGLI)

11. State two results of over-exposure of the skin to infra-red radiation. (CGLI)

18

BASIC CHEMISTRY

AN INTRODUCTION TO CHEMISTRY

What is chemistry?

Chemistry is concerned with changing the groupings of atoms, in other words, making new molecules from old. Chemicals for cosmetic and medical uses can be produced from materials of animal, vegetable or mineral origin, and foods can be changed within the human body into glucose, amino acids and glycerol plus fatty acids, which are essential to life. Some familiar chemicals are shown in Figs. 18.1 and 18.2.

As an example of a *chemical change* we can consider making zinc oxide (ZnO) from zinc metal (Zn) and oxygen (O_2). This is illustrated in Fig. 3.8 (p. 20). This zinc oxide is a white powder used in face powders, dusting powders, and in ointments. It offers a mild astringent action as well as being soothing and protective.

(a)

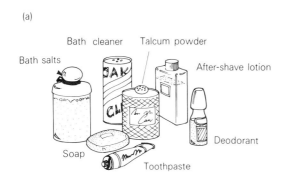

(b)

Fig. 18.1   Chemicals: (a) in the bathroom;   (b) in the kitchen

Chemical reactions

Consider still the example of ZnO. This can be made by heating zinc in air so that it combines with oxygen from the air. The molecules have been brought together to form ZnO molecules and the process is called a *chemical reaction*.

The reaction can be represented by a chemical equation as follows:

Zn + O = ZnO

In such equations the chemicals that we begin with (the *reactants*) are written on the left-hand side of the equals sign; the new chemicals (*products*) that result from the reaction are written on the right.

Since oxygen atoms are not found singly in air but only as diatomic molecules (O_2) the equation should be written with O_2 on the left; this molecule will react with two Zn atoms and produce two ZnO molecules.

$$2Zn + O_2 = 2ZnO$$

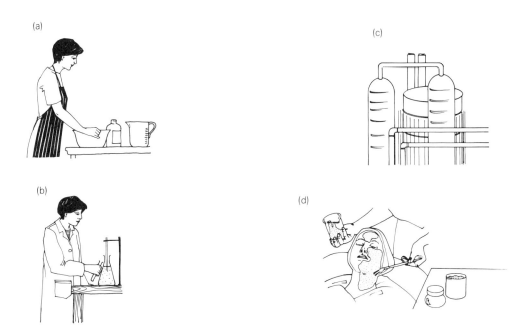

Fig. 18.2 Chemistry: (a) at home; (b) in the laboratory; (c) in industry; (d) in the clinic

The effect of temperature on a chemical reaction

Heating helps chemical reactions to get started and, unless the reaction produces heat itself, it may be necessary to continue heating the reactants in order to obtain a useful amount of product in a reasonably short time.*

If the reaction produces molecules with smaller chemical energy, then we do expect heat or other energy to be produced.†

---

*This is to be expected because heating means making the atoms move about more with the result that they will more quickly meet new partners.

†A reaction that produces heat (as well as the expected products) is described as *exothermic*. A reaction where heat energy is taken in and becomes chemical energy in the products is called an *endothermic* reaction.

Some exothermic reactions produce light as well as heat, i.e. they burn with a flame or glow. In the above-mentioned example of Zn combining with oxygen, heat and light are produced, the Zn burning with a greenish white flame. The zinc does, however, need to be heated to start off this exothermic reaction.

Oxidation

When a substance is made to combine with oxygen the process can be called an *oxidation*. Removal of oxygen is called *reduction*.

For example, in the equation $2Zn + O_2 = 2ZnO$, zinc is oxidised, but heating ZnO with carbon (C) gives the reaction $ZnO + C = Zn + CO$, which shows that ZnO is reduced while carbon is oxidised.

Many oxides can be reduced by making them react with hydrogen, because hydrogen has a great affinity for oxygen (it is very 'keen' to form water, H_2O).

A chemical that is good at giving oxygen to other reactants is called an *oxidising agent*. A chemical that is good at taking oxygen from another chemical is called a *reducing agent*. Thus oxygen and ZnO were oxidising agents in our examples and carbon and hydrogen were reducing agents.

Rusting of iron, producing the well-known brown oxide of iron, is an example of oxidation. Water plays an important part in the reaction.*

Bleaches

Many *bleaches* achieve their action by oxidising or reducing the chemicals responsible for the unwanted colour. An aqueous solution (i.e. in water) of sodium hypochlorite (NaOCl) is often used in domestic bleaches, and hydrogen peroxide (H_2O_2) is commonly used for hair bleaching. (Fig. 18.3).

Disinfectants frequently work because they are oxidising or reducing agents (Fig. 18.4), and these are described in Chapter 22.

A further example of oxidation is the production of melanin when sun-tanning occurs.

Finally, the oxidation of food to produce the energy we need is described in Chapter 19.

Hydrogen peroxide solution

Dilute ammonia solution

Fig. 18.3 Bleaching hair with hydrogen peroxide (an oxidising agent)

Fig. 18.4 A chemical may act as a bleach and as a disinfectant

For lightening the colour of the skin, instead of using a make-up containing a white powder such as titanium dioxide (TiO_2) or ZnO, a bleaching cream may be used. The active ingredient in such a cream can be hydrogen peroxide (on oxidising agent) hydroquinone. A compound of ammonia and mercury has also been used. This last bleaching agent is poisonous: both this, and hydroquinone preparations, require great care in use and can cause skin irritation in some cases.

---

*The ideas of oxidation and reduction are usually extended so that, for example, the addition of any 'electron-seeking' element to a chemical is also called *oxidation* and its removal *reduction*, but this is not important here.

Chemical radicals

(Chemical radicals are explained in the footnote for those who are interested.*)

A little more about chemical reactions

(Some further information is given in the footnote.†)

Fig. 18.5 illustrates a chemical reaction in which a solid product is obtained from liquid reactants — the product is a *precipitate*.

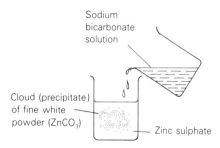

Fig. 18.5 Precipitating zinc carbonate

Acids and alkalis

An *acid* is a substance which, dissolved in water, gives a lot of hydrogen ions, i.e. there will be more H^+ than OH^- ions compared with the situation in pure water.‡

An *alkali* is the opposite to an acid because it is a substance that is dissolved and gives an excess of OH^- ions.

Some acids and alkalis are very corrosive, meaning that they cause damage to, or 'burn', many materials including skin. This possibility must always be remembered. Acids that can be tasted mostly have a sour taste.

Examples of acids include *hydrochloric acid* which is produced in the stomach and is essential for digestion of

---

*A chemical radical is a group of atoms which does not exist normally on its own but is found as part of various molecules. Examples of such groups are the carbonate group CO_3 which occurs in zinc carbonate ($ZnCO_3$), used in calamine lotion, and sulphate (SO_4) as an aluminium sulphate ($Al_2(SO_4)_3$), used as an antiperspirant. Ammonium (NH_4) and hydroxide (OH) radicals are found in ammonium hydroxide (NH_4OH). The bicarbonate radical HCO_3 occurs in calcium bicarbonate $Ca(HCO_3)_2$ which is one of the causes of hardness of water, while sodium bicarbonate ($NaHCO_3$) is well known as cooking soda.

†If some zinc sulphate solution is mixed with some sodium hydrogen carbonate (sodium bicarbonate) solution, an insoluble white powder comes from the solution (i.e. is precipitated) and bubbles of carbon dioxide gas escape. Some water is also added to that already present. The powder is zinc carbonate which, as mentioned earlier, is often used in calamine lotion that is so useful for reducing skin irritations. This chemical reaction can be represented by the following equation.

$$ZnSO_4 \quad + \quad 2NaHCO_3 \quad = \quad ZnCO_3 \quad + \quad Na_2SO_4 \quad + \quad CO_2 \quad + \quad H_2O$$

| Zinc sulphate | Sodium bicarbonate | Zinc carbonate | Sodium sulphate | Carbon dioxide | Water |

Note that the correctness of the equation can be confirmed by checking that the number of each type of atom is the same on the left as on the right of the equation.

It should also be noticed that chemical reactions normally require at least one of the reactants to be liquid or in solution or to be a gas. This is because only in fluids can molecules move about easily and so ensure a continual bringing together of the reacting molecules.

‡It should be pointed out that H^+ ions normally hold on to water molecules and so exist as H_3O^+ ions, known as hydroxonium ions.

food. This acid is also produced during epilation by the galvanic method. *Boric acid* is an antiseptic used sometimes in talcum powders. *Acetic acid* is found in vinegar, *citric acid* in lemons, and so-called *fatty acids* are used to make soaps and feature in cosmetics recipes.

An example of an alkali is *sodium hydroxide (caustic soda)* which is produced during galvanic epilation and is responsible for the destruction of the hair root. *Potassium hydroxide* is a similar alkali.

The warning sign for a corrosive chemical is illustrated in Fig. 18.6.

Fig. 18.6 The warning sign for a corrosive chemical

pH of a solution

The *pH* is a number that describes whether a solution is acid, neutral or alkaline. A pH value of 7 means neutral, 0 up to 7 indicates acid, with the acidity (i.e. concentration of H^+ ions decreasing with the pH number), while above 7 indicates an alkaline solution. Skin is normally acid with a pH of 5 or 6.

Chemicals applied to the skin should have neither too high nor too low a pH if irritation is to be avoided. Antiperspirants must not be too acid and bath salts must not be too alkaline. Too much alkali could be left in a soap.

Neutralisation of acid by alkali or vice versa

When acid and alkali are mixed, the OH^- ions from the alkali combine with the H^+ ions of the acid to form water; thus the acid and alkali tend to cancel or neutralise each other.

Examples of *neutralisation* include the neutralisation of a fatty acid by an alkali such as sodium hydroxide to produce a soap, and the neutralisation of beeswax with borax to produce a similar result.

When hydrogen peroxide is used for bleaching skin it is usual to add to it an alkali (ammonia solution, for example) in order to neutralise some H^+ ions in the peroxide solution. This encourages more ionisation of the peroxide (giving H^+ and HO_2^- ions), producing more HO_2^- ions which are responsible for the bleaching.

Salts

These are compounds that result from the neutralisation of an acid by an alkali and their solutions are ionised to some extent. Common salt is sodium chloride (NaCl) and it is formed by the reaction of hydrochloric acid and sodium hydroxide.

$$HCl + NaOH \rightarrow NaCl + H_2O$$

Sodium chloride in solution produces Na^+ and Cl^- ions and conducts electricity (as explained in Chapter 5).

ORGANIC CHEMISTRY

Organic materials*

Organic chemicals are essential in the construction of all plants and animals. They are found, therefore, in vegetables and vegetable oils and in fruits, in materials like cotton, and in fossil fuels like coal, petroleum oil and natural gas. Animal products include foods, oils and wool.

Terminology of organic chemistry

It is unreasonable to expect a beauty therapist to have considerable knowledge of *organic chemistry*, but those who are interested may like to read the footnotes in this chapter. [†]

---

*Chemicals which are described as organic consist of molecules built up of various atoms (there are several possibilities), frequently hydrogen, attached to carbon atoms — the carbon atoms sometimes being many in number and linked to form a framework. A typical organic molecule is illustrated in Fig. 18.7 (glucose):

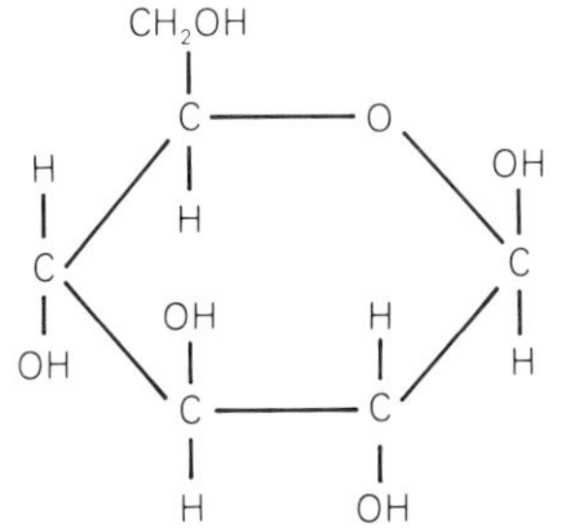

Fig. 18.7   The structure of an organic molecule — a glucose molecule

Carbon atoms are able to link so easily with each other and other atoms as well because they are tetravalent (four bonds each) and their bonds are strong.

† The simplest organic substances (*alkanes*) consist of carbon and hydrogen atoms only and their names indicate the number of carbon atoms in each molecule. Examples are shown in Fig. 18.8.

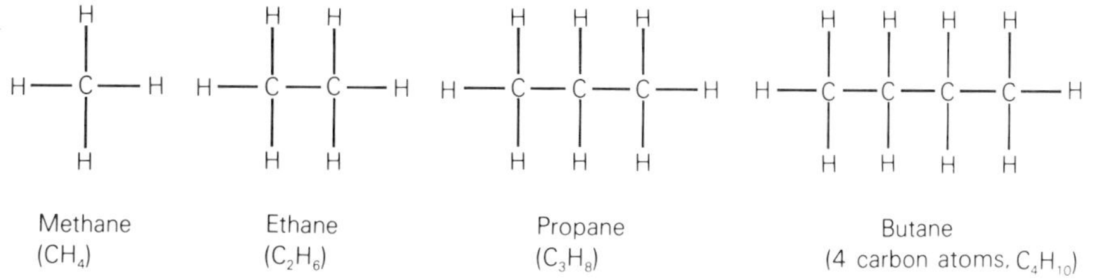

Methane (CH_4)   Ethane (C_2H_6)   Propane (C_3H_8)   Butane (4 carbon atoms, C_4H_{10})

Fig. 18.8   Some simple organic molecules

In some organic compounds a double bond occurs between two of the carbon atoms. One example is ethylene (C_2H_4) or ethene as shown in Fig. 18.9. Another example is propylene. Triple bonds can even occur in some organic molecules. Organic materials with double or triple bonds are described as *unsaturated*.

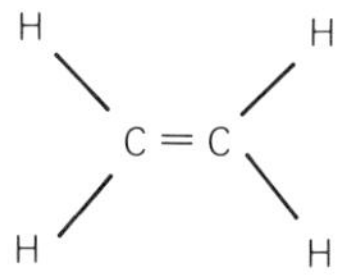

Fig. 18.9   The structure of ethylene (C_2H_4)

Alcohols

These are organic compounds in which an OH group replaces a hydrogen atom in an alkane. An important example is *ethyl alcohol* (properly called *ethanol*) where an OH replaces an H atom in an ethane molecule, as shown in Fig. 18.10, giving C_2H_5OH:

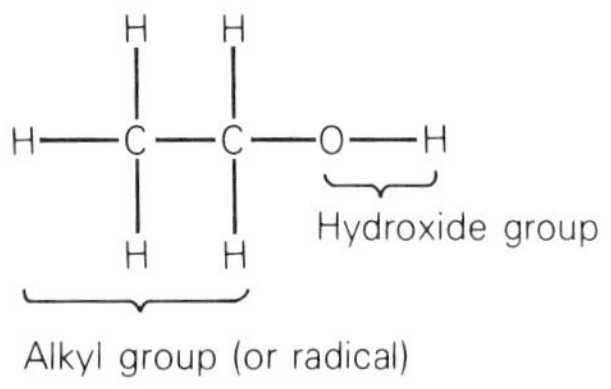

Fig. 18.10   The structure of ethanol

Ethanol is used as a solvent in the manufacture of perfumes and hair tonics, and in pre-shave and after-shave lotions. It is often called *toilet spirit*. It is also the alcohol which is drunk in wines, beers, ciders and spirits.

Methyl alcohol (methanol) is CH_3OH. It is poisonous and is mixed with ethanol in methylated spirits.

Glycerol (also called *glycerine*) is a compound with three OH groups in each molecule. It is a useful *humectant* (it holds moisture) employed in many cosmetics.

Organic acids; Esters; Modern names of chemicals; Aldehydes, ketones, amines, terpenes

(These topics may interest the beauty therapist, but it is not essential to know about them in detail. Interested readers should refer to the footnote*.)

Hydrocarbons

Organic compounds made up of hydrogen and carbon atoms only, of which alkanes are an example, are called *hydrocarbons*.

Mineral oil, paraffin wax and petroleum jelly are hydrocarbons.

---

*Organic acids

These mostly contain COOH groups from which the hydrogen atom quite easily becomes free to give H^+ ions in solution. Acetic acid has the formula CH_3COOH and is found in vinegar. Its modern name is *ethanoic* acid.

Fatty acids are organic compounds with a single COOH group in each molecule, these molecules being in some cases very large. *Palmitic, oleic* and *stearic* acids are examples.

Stearic acid, a slippery solid derived from plants or animals, has the formula $C_{17}H_{35}COOH$. Such large molecules are not characterised by the sour taste normally associated with acids.

Common oils like olive oil contain more than one kind of fatty acid.

Esters

These compounds are produced by the action of an alcohol on an acid. An example is ethyl acetate formed by the reaction of ethanol and acetic acid.

$$C_2H_5OH + CH_3COOH \rightarrow \underset{\substack{\text{acetate} \\ \text{group}}}{CH_3COO} \; \underset{\substack{\text{ethyl} \\ \text{group}}}{C_2H_5} + H_2O$$

Ethyl acetate is a liquid with a fruity smell and is found in fruits. It is used as a solvent in nail varnishes and nail varnish removers and in the manufacture of perfumes.

Oils, fats and waxes

These are characterised by their water repellency and insolubility in water; they may be lipids, hydrocarbon oils or essential oils.

Most common fats and oils that can be called *lipids* are esters formed from glycerol and fatty acids. The term *fat* is usually employed when the substance is solid at room temperature, and *oil* when it is liquid at room temperature. The former are usually animal products while the latter are usually vegetable products.

Hydrocarbon oils include petroleum, used as motor fuel, and the mineral oils used in cosmetics.

Waxes have larger molecules than fats and oils, and most are esters formed from fatty acids and alcohols simpler than glycerol. They are used in cosmetic manufacture, particularly for thickening liquids, and are of interest in connection with wax treatments (as explained in Chapter 21).

Aromatic compounds

Aromatic compounds obtain their names from the fragrant odours associated with some of them. Examples of such compounds are clove oil and almond oil.*

Modern names of chemicals

A system has now been introduced whereby chemical names are used that describe the molecular structure of the compound concerned. These names are the *systematic* names as opposed to the *common* names of the chemicals which are mostly used in this book.

Methyl alcohol, ethyl alcohol, acetic acid, ethylene, ethyl acetate and *glycerol* are common names. The systematic names are, respectively, *methanol, ethanol, ethanoic acid, ethene, ethyl ethanoate* and *propanetriol.*

Meth tells us 'one carbon atom', *eth* implies 'two carbon atoms', *an* or *ane* means 'no double bonds' but *ene* tells us that a C = C bond is present, *ol* means 'an alcohol with an OH group', while *triol* indicates 'three OH groups'. A COOH group is indicated by *oic.*

The ending *yl* implies a radical such as CH_3 (methyl) rather than the complete group CH_4 (methane), and the ending *ate* is for a salt formed from an acid whose name ends in *ic.*

Numbers may also appear in a systematic name to indicate to which carbon atoms the groups are attached. For example, *1,2,3-propanetriol* means an 'organic molecule with three carbon atoms, no double bonds, and three OH groups attached to the first, the second and the third carbon atoms'. This is, in fact, glycerol again!

Table 4, p. 293 lists the systematic names of a number of important chemicals.

Aldehydes, ketones, amines, terpenes

These are examples of other classes of organic chemicals. Aldehydes and ketones contain a carbonyl CO group linked to an organic group and a hydrogen atom in the case of aldehydes, or to two organic groups in the case of ketones. Acetone (or propanone) is a ketone. It is a highly flammable liquid and is used in nail varnishes and removers. Amines consist of an ammonia (NH_3) molecule with one of its H atoms replaced by an organic group. Thus an amine and any compound described as amino contains an NH_2 group. Terpenes are compounds formed from C_5H_8 groups.

*The molecular structure of aromatic compounds includes one or more rings of six carbon atoms. Examples are shown in Fig. 18.11.

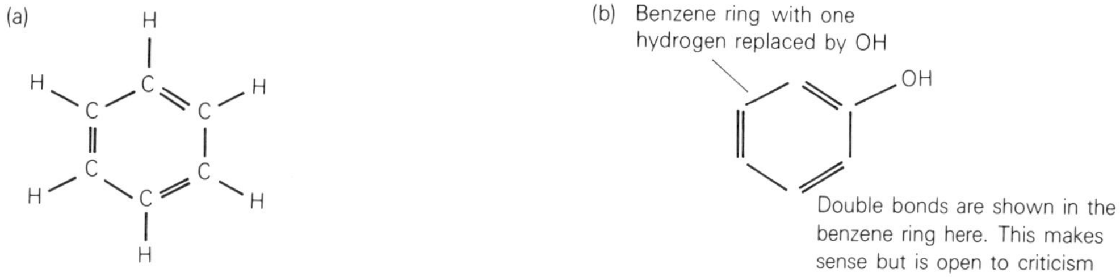

Fig. 18.11   Aromatic compounds: (a) benzene;   (b) phenol

Essential oils

Such oils are volatile and odorous, soluble in alcohol but not very soluble in water. They are very important in perfumery.* They are discussed further in Chapter 21.

Polymers

To most people *polymers* means 'plastics and man-made fibres', from plastic kitchenware, to the case of a transistor radio, to nylon stockings and other synthetic materials used for blouses, skirts, etc.

Some polymers occur in nature. Starch is an important ingredient of many foods and is an example of this; cellulose is another. These are discussed in Chapter 19.

As the name implies, the molecule of a polymer is made up of many (*poly*) parts linked together, each part being called a *monomer*. For example, polythene has the construction shown in Fig. 18.12(a) and is made by joining up many ethylene molecules — see Fig. 18.12(b).

Polyvinyl chloride (PVC) is similar but contains some Cl atoms.

(a)

(A very long chain)

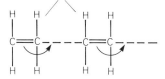

(The double bond is broken and the spare bond is used to link the monomers, as shown by the arrows)

Fig. 18.12   (a) Polythene;   (b) how it is made from ethylene molecules

Resins

These are organic polymers that may be naturally occurring or synthetic. Natural *resins* are commonly obtained when certain plants are injured; a fluid oozes from the wound and hardens into a solid resin. These resins are often odorous and useful in perfumery.

Balsams are mixtures of resins with essential oils and can be found naturally. Some balsams are used in perfumery because of their fragrance. Myrrh is an example.

Thermoplastics

These are polymers that can be softened or melted by warming. PVC and polythene are in this category.

Thermosetting plastics

These cannot be softened or melted by heating.[†]

---

*Chemically, *essential oils* are mixtures of terpenes, esters, aldehydes, alcohols, ketones, etc. These chemicals have an irritant effect on the skin and when taken internally. Caution is therefore advised in their use, as with most chemicals.

[†]Here the molecules are not simply large separate chains; instead, bonds exist between the chains as they lie side by side. Heat decomposes the whole polymer before significant flow can occur. These plastics are more rigid and are used for instrument cases, door handles, etc.

Expanded polystyrene and polyurethane foam are polymers with air or carbon dioxide bubbles forced into them.

Soaps

These are sodium and potassium salts of stearic, palmitic and oleic acids. They are made by boiling the fat or vegetable oil with a sodium or potassium hydroxide solution. The products of the reaction are glycerol and the required *soap* (sodium stearate, potassium stearate, sodium oleate, palmitate, etc.).

The properties of soaps are also affected by the amount of glycerol that remains in with the soap when it is separated from the water and glycerol in the reaction vessel.

Softer soaps are preferred for toilet soaps.

Hardness of water

Most rain falls upon open land so that it will pass through soil and rocks eventually reaching rivers from which reservoirs are filled. Consequently, the water supplied to our homes and to the salon has been in contact with minerals such as limestone rocks from which chemicals dissolve into the water. Some of these chemicals have the effect of destroying soap, and creating scum in our washing water. Water is said to be *hard* when it contains chemicals causing this problem.

So-called *temporary hardness* is due to calcium hydrogen carbonate (calcium bicarbonate) or magnesium hydrogen carbonate (magnesium bicarbonate) dissolved in the water. This hardness is called temporary because it can be removed by boiling the water. Temporary hardness is expected when rain water (containing a little carbon dioxide making it slightly acid) flows through limestone (calcium carbonate) rocks.

Permanent hardness occurs when chlorides or sulphates of calcium or magnesium become dissolved, in small concentration, in water. Such hardness cannot be removed by boiling.

Detergents (discussed in Chapter 22) are less affected by water hardness.

To deal with hardness we could distil the water before use, but obviously this could only be used for very small quantities of water; boiling to remove temporary hardness is not very practicable either. Adding sodium carbonate (washing soda) to the water is far more useful. The resulting chemical reaction produces a precipitate of calcium and magnesium carbonates leaving sodium sulphate, chloride or bicarbonate in solution. These cannot react with the soap.

Bath crystals can be sodium carbonate crystals, usually with some colouring added.

Another treatment for hard water is to pass it slowly through granules of a chemical (such as 'Permutit', which is sodium aluminium silicate) that removes calcium and magnesium ions from the water and replaces them by sodium, just like washing soda does. With such a water softener it is necessary from time to time to send sodium chloride solution through the Permutit to regenerate it (i.e. put sodium ions back into it). The apparatus may see to this automatically so that all the user has to do is to put in a quantity of common salt when required.

An alternative is an ion-exchange water purifier in which water is made to flow through two resins, one of which replaces calcium and magnesium ions by hydrogen ions, while the other replaces sulphate and bicarbonate ions by hydroxyl ions. Water is therefore produced in place of the offending minerals and the resulting water is very pure. It is deionised. A purifier of this type is illustrated in Fig. 18.13.

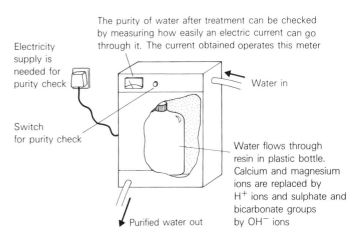

The purity of water after treatment can be checked by measuring how easily an electric current can go through it. The current obtained operates this meter

Electricity supply is needed for purity check

Switch for purity check

Water in

Water flows through resin in plastic bottle. Calcium and magnesium ions are replaced by H^+ ions and sulphate and bicarbonate groups by OH^- ions

Purified water out

Fig. 18.13 A water purifier suitable for wall-mounting

Safety with chemicals

All chemicals must be treated with respect. Many are quite harmless when used carefully and in moderation. Few, if any, are absolutely safe under all circumstances. Suppliers of chemicals are expected to point out possible hazards with labels on bottles including the appropriate information. (See Fig. 18.14.) Nevertheless, even the slightest doubt about the safeness of a chemical should tell the beauty therapist to check by reference to chemical safety books in libraries or by obtaining written guidance from the suppliers.

Is the chemical corrosive, irritant, flammable, explosive,

likely to cause cancer, a possible cause of allergies or dermatitis, or is it poisonous? These questions must be answered before use.

Furthermore, some operations, such as heating, mixing or shaking of chemicals can be hazardous. Again, risks must be anticipated. Gloves, aprons, and safety spectacles must be considered as a means of personal protection. Planning should be made in advance in case of accidents and a first aid box with eye-washing equipment should be provided.

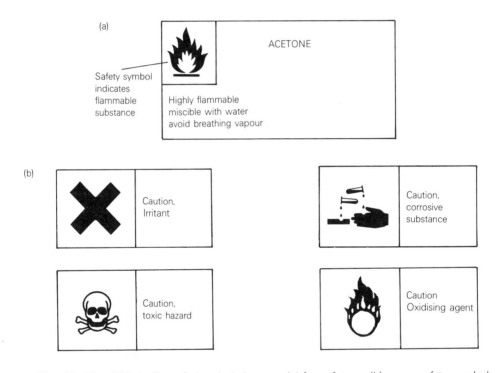

Fig. 18.14 (a) Labelling of chemicals is essential for safety; (b) some safety symbols

Fire

Paper, alcohol and natural gas, for example, will readily react with oxygen of the air to produce heat and light. The reaction is called *combustion* (or *burning*) and requires that some heating, such as from a match, is first provided. The heating is needed to produce a sufficient rate of reaction and, in the case of liquids or solids also to release vapour because liquids and solids usually do

not themselves burn. Once burning has started, the heat released keeps the process going. Flames are burning gases. If the burning continues without help, or even spreads, then we have a *fire*.

A *flammable* material is one which burns with flames that will spread. A highly flammable material produces enough vapour before any heating is applied so that instant burning can be started very easily with a tiny flame or spark.

Fires, obviously, can be the cause of severe damage to the clinic and even lead to collapse of the building. The dangers to people are not only burns (and accompanying shock), and injury from falling debris, but also poisoning from the products of the combustion (such as carbon monoxide, CO). When rooms quickly become filled with smoke in place of fresh air suffocation is a particular danger.

A fire can be stopped by removing the combustible substance, by keeping oxygen from it, or by keeping the temperature down. Water puts out a fire because the steam produced keeps air from the fire. The fire is also cooled. However, water must *not* be used where it can lead to shock from electricity. Fire blankets, sand, or foam from an extinguisher, can put out fires by smothering the combustible material so that air can no longer reach it. Carbon dioxide gas is blown from some extinguishers on to the fire in such a way that flames are pushed off the burning material; a cloud of the heavy carbon dioxide gas then forms over the material and excludes the air.

SUMMARY

- Chemistry is concerned with changing the way in which atoms are grouped together to form molecules. The changes are called chemical reactions.

- Oxidations are reactions in which atoms of oxygen (or other 'electron-seeking' elements) are made to combine with other atoms.

- Examples of oxidations are the formation of zinc oxide from zinc and oxygen, the rusting of iron, the bleaching of hair and textiles, and the browning of melanin to produce sun-tan.

- Acid solutions have a high concentration of hydrogen ions.

- Alkaline solutions have a high concentration of hydroxyl ions.

- Acids neutralise alkalis and vice versa.

- Acids and alkalis can both be corrosive.

- The acidity or alkalinity of a solution can be described by its pH value. Below 7 is acid and above 7 is alkaline.

- The normal skin is slightly acid.

- Salts are the product of neutralising an acid with an alkali.

- Organic materials contain carbon atoms and, usually, hydrogen atoms. They are found in plant and animal materials.

- Ethanol (ethyl alcohol) has the formula C_2H_5OH. It is an organic chemical and it is the OH that makes it an alcohol. It is found in wines, beer and spirits. It is the main ingredient in toilet spirit used for cosmetics.

- Glycerol is an alcohol.

- Oils are organic materials and may be hydrocarbons like petrol, lipids like animal fats and vegetable oils, or essential oils that are useful for perfumery because they are volatile and odorous.

- Polymers are materials that have long molecules.

- Resins are organic polymers. Some natural resins are odorous and are useful in perfumery.

- Soaps are formed by reacting fatty acids with an alkali.

- Hard water is water with dissolved impurities that react with soap and produce undesirable scum.

- Temporary hardness is usually due to calcium bicarbonate and can be removed by boiling.

- Permanent hardness is due to sulphates and chlorides of calcium and magnesium.

- Hardness can be removed by treatment with sodium carbonate (washing soda), or by the use of water softeners.

- A flammable material will burn and its flames can spread.

- A highly flammable material produces enough vapour to ignite very easily.

1. Explain what is meant by oxidation and give an example.

2. Name two uses of oxidation.

3. Name the atoms that make up the following molecules:

(a) common salt, (b) cooking soda, (c) washing soda.

4. What is meant by the pH of a solution?

5. Explain the meaning of the term *neutralisation of an acid.*

6. Name one hazard associated with each of the following:

(a) a strong acid, (b) acetone, (c) caustic soda.

7. Name the acid found in the stomach.

8. What are the reagents required for producing a soap?

9. What is the common name for *ethanol*?

10. What is another name for *toilet spirit*?

11. Name two polymers.

12. What property in particular is associated with essential oils?

13. Name a chemical responsible for:

(a) temporary hardness of water, (b) permanent hardness.

14. Name *two* compounds used to remove hardness from water. (CGLI)

15. What are the accepted warning signs for: (a) an irritant chemical, (b) a corrosive chemical, (c) an oxidising chemical, (d) a toxic substance?

19

NUTRITION

FOODS FOR GROWTH AND ENERGY

Food

Our bodies need fuel to provide energy for us to be able to move around, for the heart to beat, for the lungs to draw in air, and for many other functions including keeping up the body temperature. Also, suitable materials are needed for the building and repair of the body's tissues. Finally, certain chemicals are needed to control and assist in all these activities. The chemicals needed for these three purposes are called *nutrients* and are provided by the solids and liquids which we eat and drink, i.e. by our *food* (Fig. 19.1).

The intake and use of food by the body is called *nutrition* and consists of three main steps. These are, first, the intake, or ingestion, of food into the body, second, digestion by which nutrients are separated and turned into a form which the body can use, and, third, the disposal of unwanted constituents of food.

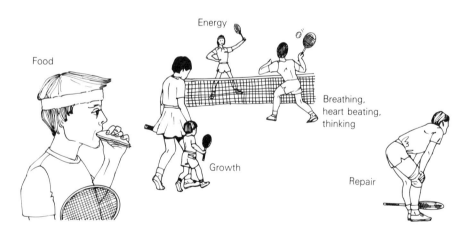

Fig. 19.1 The need for food

Classification of nutrients

The nutrients that our food should contain for health are *carbohydrates, fats* (*lipids*), *proteins, mineral* (i.e. *inorganic*) *salts,* and *vitamins. Water* is an essential requirement but it is not usually described as a nutrient.

The first three in this list are particularly important as these provide energy and nearly all the materials needed for growth and repair.

Carbohydrates

These are the sugars and starches. Molecules of *carbohydrates* are made up of carbon, hydrogen and oxygen, with twice as many hydrogen atoms as oxygen atoms in each molecule. One of the simple sugars is *glucose,* with a chemical formula $C_6H_{12}O_6$. This sugar is found in plants and fruit juices and is especially abundant in grapes and onions. It is also found in smaller amounts in the blood of all animals. Fig. 19.2 shows some common carbohydrate-containing foods.

Glucose is described as a *monosaccharide sugar* (i.e. a 'single sugar') because its ring-shaped molecule is not joined to similar molecules but remains alone. *Fructose* is another monosaccharide sugar also found in fruits. Honey is a particularly good source of fructose.

The chemical structure of alpha-glucose is depicted as shown in Fig. 19.3.

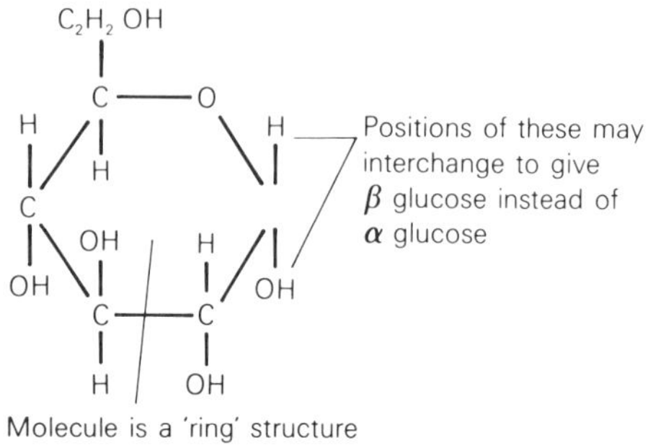

Fig. 19.2   Some foods containing carbohydrates

Fig. 19.3   The alpha-glucose molecule

Energy release from glucose

Oxidation of glucose, described by the equation below, enables the atoms concerned to achieve a lower potential energy and, in agreement with the *law of conservation of energy,* energy becomes available in some other form:

$$C_6H_{12}O_6 + 6O_2 = 6CO_2 + 6H_2O$$

It is from the oxidation of glucose that the living cells of plants and animals (including man) typically obtain energy for growth, movement and other work. This release of energy within the cell is often called *tissue respiration* or internal respiration (to distinguish it from external respiration which is the breathing of oxygen into the lungs). It should be noted, however, that this process is not as simple as the above equation suggests. In fact a series of reactions occurs, leading finally to some of the energy from glucose being given to a substance called adenosine tri-phosphate (ATP) which is found in all living cells.*

Catalysts and enzymes

Some of the chemical reactions in the body would proceed at a slow, or even negligible rate, but for the presence of *catalysts*. A catalyst is a chemical that increases the rate of a reaction without itself undergoing any overall change. *Enzymes* are organic catalysts, mostly made of proteins.

Enzymes are of vital importance in digestion. The oxidation of glucose requires enzymes to help the reactions. Even if high temperatures were possible within the human body, no appreciable glucose oxidation would occur without enzymes.

---

*Molecules of ATP have a high chemical potential energy. They are produced by conversion of adenosine diphosphate (ADP) in several of the chemical reactions that are involved in the release of energy from nutrients such as glucose. In fact the energy obtained from one glucose molecule becomes the energy of 38 ATP molecules produced in the series of reactions that convert the glucose to carbon dioxide and water. These ATP molecules act as a very successful temporary store of energy and are able to give up this energy (changing back to ADP) to tissues that need energy to work. This is especially important in the voluntary muscles of the body.

Enzymes themselves may require assistance of other chemicals, called *coenzymes,* to carry out their functions.

Disaccharide sugars

A molecule of a *disaccharide* sugar is formed from two monosaccharide molecules. For example, cane sugar is the disaccharide called *sucrose* which is formed from glucose and fructose.

Glucose + Fructose → Sucrose + Water

Sucrose is found in many popular foods (Fig. 19.4).

Lactose and *maltose* are also disaccharide sugars and are found in milk and beer respectively.

The formation of the *maltose* molecule, from two glucose molecules, is shown in Fig. 19.5.

Fig. 19.4   Sucrose is contained in many popular foods

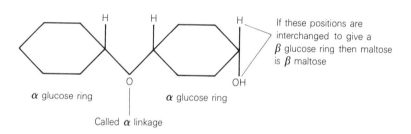

Fig. 19.5   The α maltose molecule

Polysaccharides

Polysaccharides are carbohydrates with molecules made up of large numbers of monosaccharide rings. They are therefore polymers, each monomer being a monosaccharide ring. The most important examples of polysaccharides are *starch* and *cellulose,* but the human body cannot digest cellulose. Plants contain large amounts of these chemicals. The starch is used as a means of energy storage, for example, in seeds (peas, beans, etc.). Cellulose forms important parts of the supporting structure of plants. In animals energy is stored in *animal starch* (*glycogen*).

Proteins

While carbohydrates and fats are readily available sources of energy, *proteins* are body-building nutrients, although they too can be used to provide energy. Our muscles are largely made of protein and so too is the muscle from animals (the lean meat) that we eat (Fig. 19.6, overleaf).

Fig. 19.6 Some protein-containing foods

Proteins are polymers with each molecule made up of 100 or more monomers. Each monomer is an amino acid.*

Peptides

A molecule made up of only a few amino acid groups linked together is called a *peptide* and, for a greater number up to 100, the term *polypeptide* is used.

Classes of proteins

Some amino acids are essential to health (must be supplied by the food eaten) because they cannot be made by the body from other nutrients. Foods that contain most of the essential amino acids in good proportions are called *first-class proteins* and include meat, milk and eggs. *Second-class proteins* contain useful amounts of some essential amino acids and include peas, beans and cereals.

Fats

This is the common name for the lipids other than waxes, as explained in Chapter 18. Triglycerides, phospholipids, steroids and terpenes together with waxes are all lipids.

---

*An example of an amino acid is glycine (Fig. 19.7):

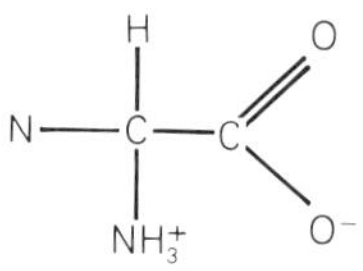

Fig. 19.7 An example of an amino acid — glycine

The presence of the NH_3 group in the molecule explains the origin of the 'amino' part of this compound's name. The presence of ions in the molecule (NH_3^+ and O^-) is responsible for the solubility of many proteins in water.

Triglycerides

Fats in foodstuffs and the fat of animals consist largely of mixtures of *triglycerides*. A triglyceride is an ester made up from glycerol combined with three fatty acids. In the human body the predominant fat is glyceryl trioleate. It is a liquid produced by the following reaction:

Glycerol + Oleic acid → Glyceryl trioleate + $3H_2O$
molecule   molecules

Triglycerides are the main chemicals used for storing energy in the body. Large deposits of such fat in tissues below the skin (subcutaneous fat) is responsible for most cases of obesity.

The main fat-containing foods are fat meats, some types of fish (such as herring), milk products and some plant seeds such as nuts and beans. Some of the sources of fat in our diets are illustrated in Fig. 19.8.

Fig. **19.8**   The fats we eat

Phospholipids

These are similar to the triglycerides but the molecules contain phosphorus atoms. These fats are found in several parts of the body. Egg yolks and liver contain phospholipid fat.

Cholesterol

Sterols are solid alcohols and *cholesterol* is a good example.*

It has been suggested that too much cholesterol in food encourages the coating of the inside of blood vessels with fat (*antherosclerosis*) and this can lead to heart attacks, strokes, etc.

Saturated fats

These are fats with no double bonds in their molecules. They are mostly animal fats and include glyceryl tristearate (tristearin) that is found in beef fat. A diet that includes a lot of *saturated fats* encourages a high level of cholesterol in the blood and therefore encourages atherosclerosis.

Essential fats

It is possible that no fats are *essential*, the body being able to synthesise the fats needed from other nutrients, but a few highly unsaturated (polyunsaturated) fatty acids do perhaps have to be provided. It is generally believed by experts that some fats are essential if tissues are not to become too dry, especially the skin. Unsaturated fatty acids, present in vegetable and fish oils, should moreover, be part of one's diet to guard against atherosclerosis.

*It has a complicated chemical structure including a four-ring group that is characteristic of a steroid.

Digestion

Food is digested in the *alimentary canal* which is really a long tube beginning at the mouth and ending at the anus, as shown in Fig. 19.9.

Digestion requires that the molecules of carbohydrates, other than the small, monosaccharide sugars of proteins and of fats, are made to undergo chemical changes to produce molecules that are small enough to pass through the walls of the alimentary canal into the body's blood and lymph. Food which is not absorbed in this way is got rid of through the anus as faeces.

The chemical changes are mostly brought about by enzymes that are put into the food as it passes through the alimentary canal. The enzymes enter the digestive system in digestive juices produced by glands.

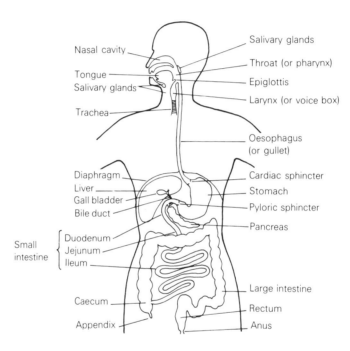

Fig. 19.9 The human digestive system — the alimentary canal

The stages of digestion

While in the mouth food is chewed to make it easy to swallow and to increase its surface area. *Saliva,* coming from the salivary glands, enters the mouth, when the taste or smell of food is received, and food makes contact with the walls of the mouth cavity. The saliva, which is largely water, contains mucus to lubricate the food, and the enzyme *ptyalin*, as well as various inorganic salts. The ptyalin is also called *salivary amylase,* and this starts the digestion of starches converting them to smaller starch-like molecules (*dextrins*) and *maltose.*

It should be noted here that, in a sense, digestion begins before food is even eaten, i.e. when the food is cooked, because cooking breaks down cellulose which would otherwise protect nutrients against being digested (see p. 218).

Food in the mouth

Food in the mouth is finally shaped into a lump called a *bolus* and is moved by the tongue to the back of the mouth. Here it is swallowed into the *gullet* while the *epiglottis* is moved to cover the windpipe so that food does not 'go down the wrong way'.

Food in the gullet

The *gullet* is a muscular tube which squeezes behind each bolus pushing it down towards the stomach. This kind of squeezing action is called *peristalsis* and quickly moves the food into the stomach via the *cardiac sphincter* (a sort of non-return entrance valve).

Food in the stomach

Food usually spends a few hours in the *stomach* which moves continually so as to churn its contents. Gastric juices containing hydrochloric acid (HCl), mucus and the enzyme *pepsin* are added to the food in the stomach, coming from glands in the stomach lining. The production of these gastric juices is stimulated by eating, and even by the sight of food, but mostly it is the hormone *gastrin* that provides the stimulation. A *hormone* is a chemical that behaves as a messenger in the body. In the case of gastrin, this is produced by the stomach lining due mainly to the chemical action of certain substances (called *extractives*) in the food. The hormone enters nearby blood vessels and the blood carries it throughout the body and thus it reaches the gastric glands.

The hydrochloric acid in the stomach stops the action of the ptyalin in the food because ptyalin cannot work in acid surroundings. This acid does, however, help to digest carbohydrates by catalysing the conversion of some sugars and starches to smaller molecules. Protein digestion also begins in the stomach by the combined action of the hydrochloric acid and the pepsin. As a result, proteins are converted to smaller molecules (*proteoses* and *peptones*). The mucus produced by the gastric glands and also by the stomach lining protects the stomach wall against abrasion and against being digested itself.

The possibility of the gastric glands being digested by their own juices is reduced, it is usually claimed, because an inactive enzyme, *pepsinogen,* is produced by the gastric glands; this changes to pepsin when it mixes with hydrochloric acid.

Food leaves the stomach as a semi-liquid fluid mixture called *chyme.* Food that softens only slowly will have spent longer in the stomach and fats delay emptying, since, as mentioned earlier, they tend to float on top of the other fluid. (Even then it empties within about five hours.)

Food in the duodenum

On entering the *duodenum,* which is the first part of the *small intestine,* the food becomes mixed with *pancreatic juice,* which comes from the pancreas gland via the pancreatic duct, and with *bile* that is produced in the liver, stored in the *gall bladder* and enters the *alimentary canal* via the *bile duct.* Also, intestinal juices enter the chyme from the walls of the duodenum.

The pancreas

The *pancreas* produces its juice as a result of its stimulation by the hormone *secretin,* produced by the duodenum lining when chyme enters from the stomach, and as a result of stimulation via nerves. The pancreatic juice contains chemicals that make it alkaline and also three enzymes, called *trypsinogen* (inactive), *amylopsin* and *steapsin*. These enzymes are more easily described as pancreatic protease, pancreatic amylase and pancreatic lipase because they are responsible for digestion of protein, carbohydrate and fat foods respectively. The pancreatic lipase converts fats into fatty acids and glycerol. The pancreatic amylase converts starches into maltose. The trypsinogen has no use at this stage. The alkalinity of the pancreatic juice reduces the acidity of the chyme so that the pH is suitable for the necessary enzyme actions.

Bile

Bile contains red and green pigments which are really waste products excreted from the liver as the result of the breakdown of old red blood corpuscles. Salts also are contained in the bile; these emulsify fats, the resulting minute globules of fat being easier to digest. Furthermore, the bile salts help the pancreatic lipase action, and, being alkaline, help to reduce the acidity of the chyme.

Intestinal juice

This juice will complete the digestive process. It is supplied by the intestinal wall of the duodenum and the small intestine beyond. It is an alkaline liquid and contains *enterokinase* to convert *trypsinogen* to the active enzyme *trypsin* as well as enzymes for digestion of peptides, disaccharide sugars and fats (i.e. peptidases, disaccharidases and intestinal lipases). Trypsin converts peptones and proteoses to polypeptides and peptides which the peptidases break down into the final amino acids. The disaccharidases complete the conversion of carbohydrates to monosaccharides like glucose, while the lipases convert fats not already converted by the pancreatic lipase into fatty acids and, of course, glycerol. An outline of the digestive processes is given in Fig. 19.10.

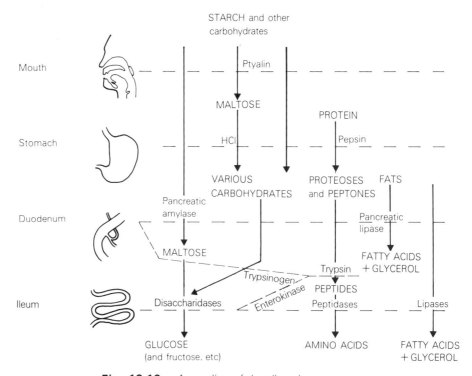

Fig. 19.10 An outline of the digestive process

Absorption of digested food

The molecules of glucose, amino acids, fatty acids and glycerol are small enough to pass through the lining of the small intestine so that they can really enter the body.

The intestine walls are very folded and are covered with hair-like projections called *villi* so that a large surface area is available for the nutrients to pass through. A to-and-fro movement of the villi also assists the mixing of the intestinal contents. Glucose and amino acids mostly enter the blood that flows through the intestinal walls

and villi, while fatty acids and glycerol mostly enter the lymph in the lymphatic vessels (Fig. 19.11). It seems that most of the fatty acids and glycerol immediately form triglyceride fats again once they are in lacteals; then they travel away in the lymph as small fat droplets in suspension and as dissolved fat, i.e. the very small drops of fat stabilised by certain low-density proteins (see Chapter 9).

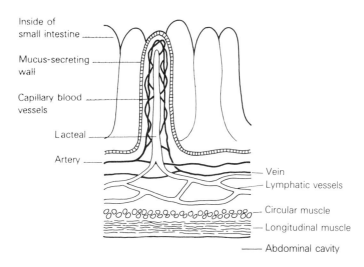

Fig. 19.11 The wall of the small intestine

In the large intestine

The undigested remains of food plus the digestive juices (largely water) enter the *large intestine* as a pulp, still slightly acid. Water passes from this mixture through the intestine wall into the blood vessels so that the contents of the intestine acquire the characteristic texture of the faeces that finally enter the rectum and leave the

alimentary canal at the anus. The undigested food is mainly *cellulose* (the roughage of the food we eat). Bacteria in the large intestine may break down cellulose to some extent but too late for the carbohydrate products to be absorbed into the body.

Obtaining energy from the absorbed nutrients

As mentioned earlier, the oxidation of glucose to water and carbon dioxide is achieved as the net result of a whole series of chemical reactions (producing molecules of ATP from ADP). These reactions can be grouped into two stages known as *glycolysis* ('break down of glucose') and the *citric acid cycle*. The first of these

stages converts glucose to *pyruvic acid* if oxygen is available but, if there is inadequate oxygen supply then the pyruvic acid is forced to change into *lactic acid*. Less ATP is formed (i.e. less energy will be made available from the original glucose) in the latter case.

If oxygen is available then the next stage can occur whereby pyruvic acid is converted to citric acid and hence to CO_2 and H_2O.

A very simplified outline of these processes is illustrated in Fig. 19.12; some other features are also shown.

One of the functions of the liver is to remove glucose from the blood and convert it to glycogen (animal starch), and store this. The process is encouraged by the hormone *insulin*. When necessary, the liver changes its glycogen into glucose so as to keep a suitable level (i.e. concentration) of glucose in the blood. The glycogen to glucose change is encouraged by hormones such as *glucagon* and *adrenalin*. The reversibility of the liver's action in this respect is clearly indicated in Fig. 19.12.

Glucose can also be converted to fats or proteins and both fats and proteins can be oxidised to provide energy.*

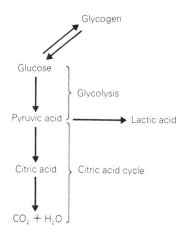

Fig. 19.12   The oxidation of glucose

The liver

The position of the *liver* in the body can be seen in Fig. 19.9 (p. 210). The most obvious functions of the liver are the production of bile, the conversion of glucose to glycogen, and the storage of this glycogen in the liver cells.

We see in Fig. 19.14 (a) that blood enters the liver from the intestine carrying any glucose that has been absorbed from the intestine, together with some amino acids. This blood, mixed with oxygenated blood that is needed for the well-being of the liver cells, flows through

*The steps by which not only glucose but also other nutrients may be oxidised are illustrated in Fig. 19.13.

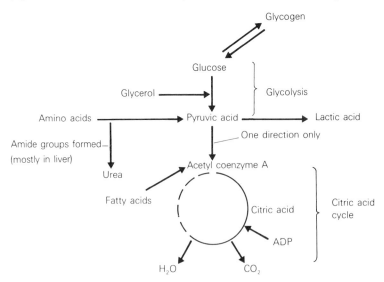

Fig. 19.13   How nutrients are oxidised

All the changes shown in the figure are reversible (although the arrows show only the direction from nutrients *to* CO_2 and H_2O), *except* that the pyruvic acid to acetyl coenzyme A conversion is in one direction only. This means that glucose can be changed to fatty acids or even to amino acids, but fatty acids cannot be changed into glucose. The dotted path shows that the sequence of reactions in the citric acid cycle can continue so as to take us back to the acetyl coenzyme A again, i.e. the reactions form a cycle.

channels called *sinusoids* (Fig. 19.14(b)) and the liver cells in contact with this blood obtain chemicals from the blood. From these chemicals the liver cells make bile which passes from the cells into bile channels, as shown in Fig. 19.14(b). The bile is collected into interlobular bile ducts which join to form a single duct. This leads from the liver, past the gall bladder where excess bile is stored and, as the common duct, enters the alimentary canal where the bile plays an important part in digestion. The route taken by the bile is illustrated both in Figs. 19.14 (b) and (c), and in Fig. 19.9 (p. 210).

When the bile is not entering the intestine it fills the gall bladder which is really a bag for storing bile.

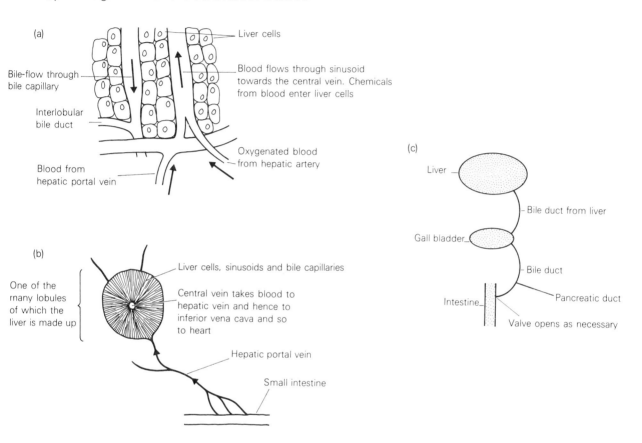

Fig. 19.14   The liver: (a) blood and bile flow; (b) a liver lobule; (c) delivery of bile to the intestine

The kidneys

The *kidneys* are excretory organs by means of which the body gets rid of unwanted water and waste chemicals, particularly nitrogen compounds such as *urea* that results from the breakdown (*catabolism*) of proteins.

The water and waste products are brought to each kidney dissolved in the blood (Fig. 19.15, overleaf). The water and small molecules pass from the blood through the walls of blood capillaries into the *Bowman's capsule*. This liquid enters the kidney tubules where about 98% of the water, and some chemicals valuable to the body, pass through the tubule walls back into the surrounding blood capillaries. The solution that remains in the kidney tubules collects in the pelvis of the kidney and is called *urine*.

From here the urine is moved by peristaltic action down the *ureter* of each kidney into the *bladder* where it is stored.

The bladder is essentially a muscular bag which, when filled sufficiently, empties by muscular action. The release of urine is called *micturition*. The action is basically a reflex action but, except in young children, voluntary control predominates.

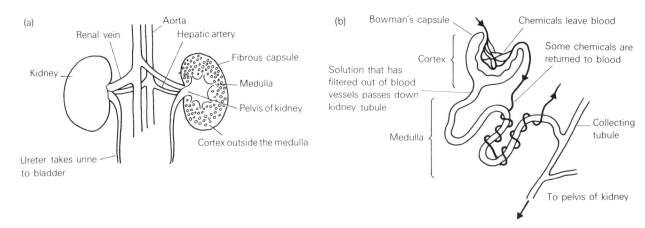

Fig. 19.15 The kidneys: (a) general features; (b) the production of urine

MINERAL SALTS AND VITAMINS

Mineral salts

In addition to the organic nutrients required by the body, it is essential that certain inorganic nutrients be provided and these, hopefully, will be contained as dissolved salts in the food and drink ingested. They are called *mineral salts* and the elements needed include *calcium, iron, sodium, potassium, magnesium, iodine* and *phosphorus*.

Calcium is needed to make bones; iron is needed for making the haemoglobin of the red blood corpuscles; sodium and potassium help to produce the correct osmotic pressure* in the body fluids and play an important role in the working of nerves; magnesium is important for many enzymes to work; iodine is required for thyroxine†; and phosphorus is needed to combine with calcium to give calcium triphosphate, which gives bones their strength and rigidity. Phosphorus, although provided in organic foods, must be supplemented by some inorganic phosphorus.

These elements, being provided as dissolved salts, enter the body as ions. For example, sodium enters the body accompanied by chlorine ions because sodium chloride (common salt) is contained dissolved in our foods.

Some of the sources of mineral salts are listed below:

Calcium (Ca) — milk
iron (Fe) — meat, egg yolk
sodium (Na) — bread, salt
potassium (K) — vegetables
magnesium (Mg) — most foods, green vegetables
iodine (I) — sea foods, most drinking water
phosphorus (P) — milk

(The symbols in brackets are the recognised symbols or abbreviations for the elements.)

Many foods such as cabbage will contain all of these in varying amounts. As Fig. 19.16 illustrates, a green

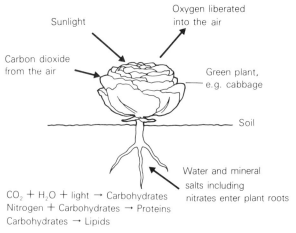

$CO_2 + H_2O + light \rightarrow$ Carbohydrates
Nitrogen + Carbohydrates $\rightarrow$ Proteins
Carbohydrates $\rightarrow$ Lipids

Fig. 19.16 Chemicals that make a plant become useful as a food

*See Chapter 9.
†See Chapter 20.

plant obtains, through its roots, water and mineral salts (including ions of Ca, Fe, Na, K, Mg, I, P and nitrogen (N) as nitrate ions) which are dissolved in the water.

Light energy is used to combine carbon dioxide obtained from the surrounding air with water to produce carbohydrates by the process of *photosynthesis*. This occurs only in the green parts of plants where *chlorophyll* acts as a catalyst.

Vitamins

Vitamins are organic chemicals that are essential to the body but are not carbohydrates, proteins or fats. They are not sources of energy but are essential catalysts. They are needed only in quite small quantities. Although not all vitamins are amines the term *vitamin* has come from 'vital amines'. Vitamins are usually described by letters A, B, C, etc., rather than by their chemical names.

Vitamin A is important for body growth and proper vision. Lack of it causes difficulty in seeing at night. Carrots and liver are good sources of vitamin A. *Vitamin B* is not really a single chemical but comprises *vitamin B₁, B₂,* etc. Lack of *thiamin* (*vitamin B₁*) causes faulty working of nerves and hence muscular weakness. Also mental confusion can result. It is found in many foods such as fresh fruit and vegetables.*

Vitamin B₂ is called *riboflavin* and is obtained from green vegetables. *Niacin* is another B vitamin found particularly in meat. Deficiency of niacin can cause dermatitis. Absence of some other B vitamins could produce *anaemia* (insufficient red corpuscles in the blood), but they are unlikely to be lacking in a person's normal food intake.

Vitamin C is *ascorbic acid* and is essential for the walls of blood vessels to be kept in a healthy condition. It is found particularly in citrus fruit (lemons, oranges, grapefruit) and also in tomatoes, most fresh fruits and fresh vegetables (i.e. uncooked). *Scurvy,* an ailment that can be fatal, is caused by deficiency of this vitamin. It was formerly common amongst sailors on long voyages.

Vitamin D is needed to ensure the proper hardening of bones and its deficiency in children is responsible for *rickets.* It is needed too for strong, healthy teeth to be formed. It is found in cod liver oil particularly, and also in fish and eggs. Besides obtaining vitamin D from foods the body can produce it when ultra-violet B radiation is reaching the body from sunlight or from a sun-lamp.

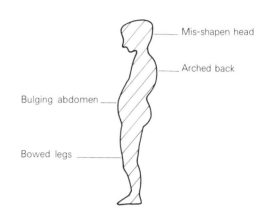

Mis-shapen head

Arched back

Bulging abdomen

Bowed legs

Fig. 19.17   The need for vitamins — the profile of a body deprived of vitamins

Vitamin E has been regarded as essential for fertility or even for maintaining a normal interest in sex, but it is no longer thought to be necessary. Vitamin E like vitamin B and D, comprises more than one chemical. It is found in green vegetables and in various cereals.

Of the above-mentioned vitamins, B and C are water soluble while A, D and E dissolve in fats. Consequently A and D are stored in body fats, but B and C must be continually supplied. It is not difficult to accumulate too much vitamin A or D, and this can be harmful to the body.

Cooking can reduce the vitamin content of foods.

---

*Vitamin B₁ seems to be necessary for the enzyme action concerned with the conversion of pyruvic acid to acetyl coenzyme A.

The cooking of foods

Quite apart from warming food, *cooking* has some important effects upon its suitability for eating.

The starch in a food is usually in the form of small lumps, called grains, each surrounded by a skin; a number of grains together are enclosed within a cellulose wall (Fig. 19.18). During proper cooking in a moist heat the grains swell and burst their skins forming a paste or jelly and the starch is said to be *gelatinised*. This jelly in turn bursts the cellulose wall. When the food is subsequently eaten, digestion of the starch is possible because the digestive juices can reach the starch. Without the cooking, digestion would be prevented because of the indigestible cellulose wall.

When meat is cooked its appearance is improved in as much as the red colour of the blood is removed. Also the fibres of the meat are loosened and the protein is coagulated, giving the meat a firmer and more easily chewed texture. In addition, the cooking can kill bacteria that might otherwise give rise to food poisoning.

Unfortunately, vitamins can be lost from food during cooking. For example, vitamin C is soluble in water and about a third of the vitamin C in a vegetable can be lost by its dissolving into the water in which it is boiled. Some of the vitamin C will also be destroyed by the action of heat.

Storage of food can also lead to loss of vitamin C content, depending upon the food concerned and upon the temperature at which it is stored.

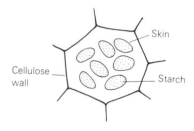

Fig. 19.18   Starch grains before cooking

THE ENERGY NEEDS OF THE BODY

The body's metabolism

Night and day our bodies are filled with countless minute movements and chemical changes, even when we are resting or asleep. All this activity is causing energy to be used so that food is oxidised, waste products such as carbon dioxide and water are formed, and heat is produced. This activity is called the *metabolism* of the body.

It is interesting to note that one perhaps unexpected activity of the body is the continual replacement of the tissues of which it is made and of the stored fat and glycogen.

Basal metabolic rate

The rate at which the body is using energy or producing heat when the person is completely at rest, so that the metabolism is as low as possible, is called the *basal metabolic rate (BMR)*. See Fig. 19.19.

The BMR is approximately 80 kilocalories per hour (340 kJ/h) or 40 kilocalories per hour for each square metre of body surface area.

It is slightly lower for women than for men and is higher for children.

The BMR is measured after the person has rested without eating for 12 hours. The room temperature should be about 20°C. The person being studied breathes in and out through an apparatus which measures how much oxygen is taken from the air by the lungs. The amount of oxygen being used tells us how much oxidation of food is occurring.

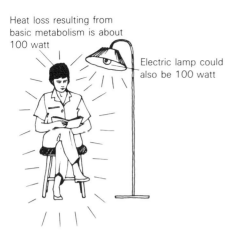

Heat loss resulting from basic metabolism is about 100 watt

Electric lamp could also be 100 watt

Fig. 19.19   The basal metabolic rate

Specific dynamic action, or thermic response

After a meal is eaten, particularly if at least mild exercise is undertaken, the metabolic rate increases so that extra energy from food is used and extra heat is produced.

This effect is called *specific dynamic action* or, better, *thermic response*.

Effect of ambient temperature on metabolic rate

In cold surroundings the metabolic rate will normally increase by perhaps 30 Calories per hour. (Remember that 1 Calorie is the same as 1 kilocalorie.)

Effect of activity on metabolic rate

The rate at which calories are used can be very much increased by activity or deliberate exercise. Heavy work can raise the calorie requirement to 3000 Calories per day or more. A three-mile walk can use up something like 300 Calories. Vigorous exercise can also produce a large thermic response lasting for some hours.

During sleep the metabolic rate is below the BMR by about 10 Calories per hour.

Calorie values of foods

(1 Calorie = 1000 calories)

| One portion of cabbage 10 Calories | Cup of white coffee 30 Calories | Portion of peas 50 Calories | Slice of bread 100 Calories | Slice of fruit cake 150 Calories | Cornflakes, sugar and milk 200 Calories | Large portion of chips 250 Calories | Pork chop 350 Calories |

Fig. 19.20   Calorie values of some common food items (given in Calories)

The maximum amount of energy that can be obtained from a gram or kilogram of food, or from an item of food, is called the *Calorie value*.

For example, complete oxidation of 1 g of glucose produces about 4 Calories using up 750 cm³ of oxygen and producing 750 cm³ of carbon dioxide. For fats and proteins the Calorie values are 9 and 5 Calories per gram respectively. A slice of bread has a Calorie value of something like 100 Calories.

Fig. 1.11 (p. 7) and Fig. 19.20 give some examples of the Calorie values of common foods.

Efficiency of energy use

A resting person needs, shall we say, 80 Calories per hour. This energy can be provided by eating a thin slice of bread for which the Calorie value is about 80 Calories. Because the body makes full use of all the available energy in carbohydrate food, the slice of bread does provide the required energy. Fats are also used with 100% efficiency but, in the case of protein, only four-fifths of the available energy is used by the body. This means that the efficiency of use is 80% while the remaining 20% of the energy is wasted and remains in the reaction products of the protein metabolism, which are excreted.

Thus, while carbohydrates provide 4 Calories per gram and fats 9 Calories per gram, the body can make use of only 4 Calories per gram of protein although the Calorie value of protein is 5 Calories per gram.

A typical daily energy requirement

The total daily need (24 hours) of a person might be as follows:

Basal metabolic rate (80 Cal per hour) for 16 hours	$= 80 \times 16 = 1280$ Cal
70 Cal per hour for 8 hours of sleep	$= 70 \times 8 = 560$
Allowance for thermic response	200
Allowance for light work	700
	———
Total:	2740 Cal

Choice of foods

Food intake must provide the necessary daily Calorie requirement of (say) 2700 Calories (or 11 MJ), unless slimming is intended.

Perhaps not more than one-fifth of this total should be eaten in the form of fats since high intake of fats (especially animal fats) encourages atherosclerosis. Protein is necessary for the building and repair of the body tissues and could form 20% of the Calorie total, with half of this being first-class protein (meat, eggs, fish, etc.). Carbohydrates can then account for the remaining 60% of the Calories needed. Fibre content is also important.

Water, vitamins and mineral salts are, of course, essential. 'Malnutrition' can mean 'eating sufficient as regards energy requirements but with an unsuitable balance of nutrients'.

Sucrose is the popular sugar for sweetening tea and coffee and is ingested with such things as cakes, biscuits and jam. This carbohydrate does little to satisfy one's appetite and possibly tends to encourage the body to produce fat. Carbohydrates also have a bad reputation regarding tooth decay.

Some foods, such as cabbage, lettuce, apples and oranges, grapefruit and cod, contain a large proportion of water and have low Calorie values so that, happily, quite large amounts of these foods can be eaten without much weight being put on. The Calorie values of some meals are shown in Fig. 19.21.

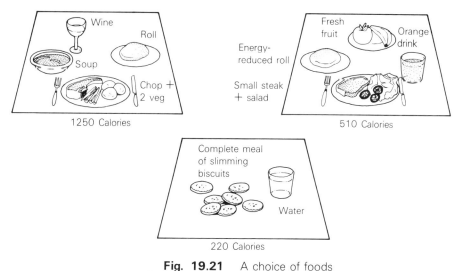

Fig. 19.21 A choice of foods

Appetite

The desire to eat is influenced by the concentrations of glucose, insulin and (free) fatty acids in the blood and is reduced when the stomach feels full. Psychological factors are important too and, of course, illness can affect *appetite*.

Obesity

If more food is eaten than is used to provide energy, then it will be stored as glycogen and fat, and the body's weight will increase as a result. An increase of body weight by 10% or more above the weight expected for the person's height and frame is regarded as *obesity* (Fig. 19.22).

More than 10% above normal weight

Fig. 19.22 Obesity

Obesity may also be caused by illness. Even when this is not the case it may be difficult to reduce weight because of the desire to eat, habit, social obligations, and frequently because of a hereditary tendency towards obesity.

Obesity can lead to health problems and to reduced life expectancy.

When an obese person's weight is steady the obesity is described as *passive*. If weight is still increasing it is called *active obesity*. During passive obesity a person may eat far less than a slim person yet lose no weight.

An obese person tends to be less active and to show a lower response to cold — the ability to burn up food, beyond what is needed for the basal metabolism and for the limited exercise undertaken, seems to be lost. Changes in the body metabolism occur, such as an increase in the concentration of insulin in the blood.

Weight reductions

For ordinary obesity it is essential to reduce the energy intake or use more energy or both.

Slimming by modification of one's diet can be achieved by careful calculation of the Calories of each item eaten (a *Calorie-controlled diet*) or by cutting out some foods (for example, all carbohydrates) without any other restrictions (a *free diet*).

Total starvation will result in a *weight reduction* of something like 2 kg per week but is not advisable. Reduction of food intake to about 1000 Calories per day is more reasonable.

Two or three meals per day are less fattening than occasional large meals even when the total Calories are the same. A meal taken just before retiring to bed is very inadvisable as the complete lack of exercise after the meal causes the thermic response to be negligible.

It is worth noting that a temporary reduction of fluid in the body occurs when carbohydrate intake is reduced, and also that fluid may accumulate noticeably in a woman's body just before menstruation. These effects should be disregarded in assessing weight reduction. A fall in blood insulin when dieting begins can, unfortunately, encourage appetite.

Slimming drugs can be useful as a slimming boost when the will to slim weakens or when all else fails. Such drugs may aim at reducing appetite, increasing the metabolism, or influencing the level of insulin in the body.

To reduce weight by exercise alone is usually difficult, but exercise is most helpful in conjunction with a reduced diet. Walking for 3 miles in 1 hour or swimming 15–30 minutes per day will typically give a weight reduction of 1 kg per month, with a further reduction expected from the accompanying increase of metabolic rate.*

Even if the energy used is small, natural exercise and the passive exercise obtainable by faradic treatment improve the working of muscles, and can improve the shape of the body and posture as a result (Fig. 19.23).

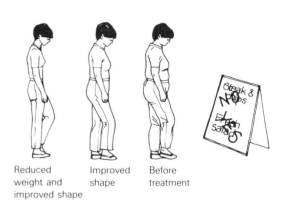

Reduced weight and improved shape   Improved shape   Before treatment

Fig. 19.23 Figure improvement

In addition to exercise, massage and galvanic treatments help the blood and lymphatic circulations and it is reasonable to suppose that these might assist the removal of stored fat during dieting.

Sauna and steam treatments (Turkish baths) provide temporary weight reduction because of the loss of water from the body by perspiration, but the body soon restores its water content to normal by retaining more of the water subsequently ingested.

SUMMARY

- The energy containing components of foods are carbohydrates, proteins and fats.

- Starch and sugars such as glucose, sucrose and fructose are examples of carbohydrates.

- Essential proteins cannot be made by the body from other food materials.

- Most animal fats and vegetable oils are triglycerides.

---

*As regards the Calories used in exercise, it is fortunate that the body has an efficiency of only about 20% for converting food energy into the exercise work. For example, to climb a flight of stairs the person's body, weighing (say) 50 kg, must be raised by about 3 metres so that the work done in this exercise is $50 \times 3 \times 10$ (see Chapter 1) or 1500 J or 0.4 Calories. Because of the 20% efficiency this means that 1500×5 or 7500 J (2 Calories) of food energy is used. The difference, namely 6000 J, becomes heat.

- Foods enter the mouth, pass down the gullet, go through the stomach, the small intestine and then the large intestine and reach the anus where the remaining waste is excreted as faeces.

- Food is mostly digested (made soluble) by enzymes.

- Carbohydrates are digested by ptyalin in the saliva, pancreatic amylase and disaccharidase.

- Proteins are digested by pepsin in the stomach, by trypsin and by peptidases in the intestines.

- Fats are digested by pancreatic lipase, and by lipases in the intestines.

- The useful products of digestion are glucose (from carbohydrates), amino acids (from proteins) and fatty acids and glycerol (from fats). These are absorbed into the body through the walls of the intestines and are carried away mostly in the blood.

- The liver produces bile and stores carbohydrate as glycogen. The kidneys remove water and waste products such as urea from the blood. The resulting urine passes to the bladder.

- Mineral salts are inorganic chemicals needed by the body but not for energy supply. Examples are calcium for bones, iron for the red corpuscles of the blood and iodine for the thyroid gland.

- Vitamins are organic chemicals required in small quantities by the body, not for energy supply: vitamin A for good vision and body growth; B vitamins for healthy skin, nerves and muscles; C for healthy blood vessels; D for healthy bones; E for fertility.

- Vitamin A is found in carrots and liver, B and C in vegetables and fruit, D in fish and eggs, E in vegetables and cereals.

- Vitamins A and D are stored in the body; we can have too much of these.

- The body metabolism is the total of all the chemical changes occurring in the body, all using energy and producing heat.

- The basal metabolism is the metabolism when we are resting.

- The calorific value of any food is the amount of energy it can provide.

- The typical daily energy requirement (for a person doing light work) is about 2700 Calories.

- A person whose weight is 10% or more above normal can be called *obese*.

- Reduction of weight requires smaller food intake or much greater expenditure of energy on exercise.

EXERCISE 19

1. Name the three kinds of energy-providing foods.

2. Name *two* mineral elements essential for health and state their functions. (CGLI)

3. What is an enzyme? (CGLI)

4. (a) Name a digestive juice secreted in the mouth.
   (b) What is its function? (CGLI)

5. For what is vitamin C important?

6. State the energy requirement for a person doing light work.

7. (a) Name the elements in a carbohydrate.
   (b) Name a food composed of pure carbohydrate. (CGLI)

8. Define the term *vitamin* giving *four* examples and a common source of each. Explain fully the importance of these vitamins by describing in detail the effects of a deficiency of each. (CGLI)

9. *Vitamin Crossword*

Across

1. The acid which is vitamin C (8)
4. Vitamin D prevents this (7)
7. One of the B vitamins (10)
8. Could a large vitamin intake make you ill? (3)
10. Another B vitamin (6)
11. These fruits are a good source of vitamin C (6)

Down

2. A good source of vitamin A (6)
3. It is caused by vitamin C deficiency (6)
5. They are provided by carbohydrates, proteins and fats but not by vitamins (8)
6. Yet another B vitamin (7)
9. You must not forget vitamins when you do this (4)

10. Write down two functions of the liver.

11. What is the main function of the kidneys?

20

ASPECTS OF ANATOMY AND PHYSIOLOGY

=========== **WHAT THE BODY IS MADE OF** ===========

——————— **The cell structure of the body** ———————

A building such as a block of flats is made up of many rooms and, in much the same way, the human body is made up of microscopic *cells*. Each cell normally consists of a blob of *protoplasm* (largely water and proteins), surrounded by a flexible wall (mostly protein and fat), and containing a denser region called the *nucleus*. Cells also contain other items such as fine threads called *fibrils* and *mitochondria*. These inclusions

vary according to the particular activities that need to take place in the various cells, and the shapes of cells vary according to their functions and locations. Fig. 20.1 illustrates a quite simple cell.

The body contains millions of cells. In some places the cells are closely packed and stuck together, as suggested in Fig. 20.2 (overleaf).

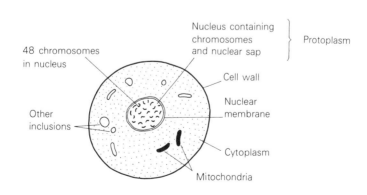

Fig. 20.1 A single cell

225

Chromosomes

In a nucleus there are *chromosomes* (Fig. 20.1), 48 in number, which provide the instructions that enable the cell to carry out its various activities. All the cells of one person have identical chromosomes in their nuclei. As a result, it is the chromosomes that determine the colour of the hair, skin and eyes and other characteristics of a person. Two chromosomes in each nucleus decide the sex of the person. Either there are two X chromosomes so that the person is female, or there is one X and one Y chromosome and the person is male.

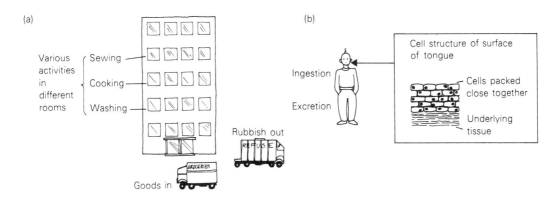

Fig. 20.2   The human body resembles a block of flats

Cell metabolism

Cells are living things, at least initially even if not throughout the whole of their useful existence. Thus cells can, and do, repair and replace their own parts. This is always necessary because the molecules of which cells are made are always breaking down. This means that chemical changes are occurring; nutrients must enter cells to supply the necessary repair materials as well as the energy required for these chemical changes. In addition to this cell *metabolism*, further chemical changes occur in many cells whereby chemicals are produced and released from the cell (i.e. are *secreted*). The walls of the cell are *semipermeable membranes* and allow nutrients, including oxygen for oxidation, to pass into the cell as well as secretions to pass out.

Cell division

In an adult, while most of the body's cells cannot be replaced once they have died or been lost in any way, there are cells (such as in the skin) which can each divide into two identical cells thus increasing the number of cells. This multiplying process is called *mitosis* and requires that each of the 48 chromosomes in the nucleus split into two identical parts and the nucleus divides into two, each new nucleus containing a complete set of chromosomes. This is seen in Fig. 20.3. Then the rest of the cell divides.

Besides mitotic division of cells, there is another kind of division that involves *meiosis* which is an essential step in the development of each *ovum* (egg cell) in the woman and each *spermatozoon* (male reproductive cell) in the

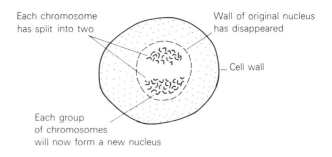

Fig. 20.3   A stage in the process of cell division by mitosis

man. It is the ovum and spermatozoon that must join together to begin the growth of a baby. In a meiotic division that produces an ovum the 48 chromosomes form two groups of 24 with an X chromosome in each group; the original cell separates into two cells with a group of 24 chromosomes in each.

Similarly, in a man, meiosis produces spermatozoons with 24 chromosomes in each; half of the spermatozoons formed have an X chromosome in them and the others have a Y chromosome.

When, after sexual intercourse, a spermatozoon fertilises an ovum, the 24 chromosomes of the spermatozoon combine with the 24 of the ovum to form a normal cell with 48 chromosomes. This cell will divide and become the baby.

If it is a spermatozoon cell with a Y chromosome that joins with the cell in the woman, the resulting cells of the baby will each have one X and one Y chromosome and the baby is a boy. If, instead, the spermatozoon cell contains an X chromosome then each cell of the baby will have two X chromosomes and the baby will be a girl.

Heredity

As already explained, the inherited characteristics are decided by the structural details of the chromosomes that are received from the parents. A person's shape, size, colour, mental ability and other characteristics are decided to a large extent by *heredity*. What happens to a child before and after birth also plays an important part.

Types of cells and tissues

Some parts of the body, for example, bones, need to be strong. Other parts, like the skin, need to be flexible. The blood is fluid, and muscles must be able to contract. Different types of cells can be used, and these can be differently connected to produce a variety of tissues to suit these requirements.

Tissues are classified as:

(a) *Epithelial tissues.* These form the skin and linings of such structures as the alimentary canal, the lungs, the blood vessels and other parts. Also included are the linings of glands where the epithelial cells secrete important enzymes, hormones or other important fluids. Fig. 20.2(b) shows the epithelial tissue which forms the surface of the tongue and Fig. 20.4 shows the structure of some glands.

(b) *Connective tissues.* Supporting the important parts of the body, keeping them in place and filling in the spaces between them are the functions of the connective tissues. The cells are well separated, lying in a jelly-like fluid, and they secrete chemicals which form long fibres which lie tangled between the cells. The fibres can be of two kinds. *Elastin fibres* are yellow and elastic (i.e. can be stretched and will return to their original size) while the other kind are strong but flexible. These white, *collagenous fibres* occur in bundles.

Areolar tissue is a loose connective tissue and the tissue beneath the skin is of this kind. *Adipose tissue* is areolar

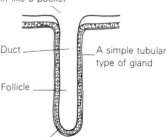

The gland is really a continuation of the epidermis which is pushed in like a pocket

Duct

A simple tubular type of gland

Follicle

The cells secrete liquid into the follicle and this liquid passes through the duct to the skin surface

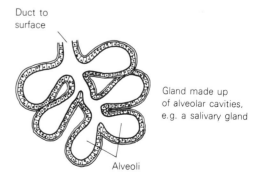

Duct to surface

Gland made up of alveolar cavities, e.g. a salivary gland

Alveoli

Fig. 20.4 The structure of some glands

tissue containing special cells that can store appreciable amounts of fat, as shown in Fig. 20.5. Cartilage and bone are also classed as connective tissues and discussed later in this Chapter.

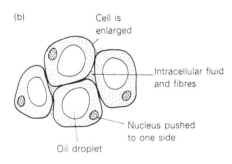

Fig. 20.5 (a) Areolar tissue; (b) adipose tissue

A condition can arise where there is an abnormally large concentration of fluid and increase in the amount of fat in the adipose tissue beneath the skin, this being accompanied by thickening of the connective fibres in the area. The tissue is then described as *cellulite* and it gives the skin a bumpy, quilted appearance, with the bloated tissue bulging out between bands of hardened fibres.

(c) *Muscle.* Muscle consists of large numbers of long, thin fibres lying alongside each other. There are two main types of muscle, *plain* (or *striated*) and *smooth* (*unstriated*). *Heart muscle* is a third kind. In the fibres of plain muscle there are numerous threads which appear striped along their length when seen under a microscope, giving rise to the term *'striated'*. In smooth muscle these threads are few in number.*

(d) *Blood.* This consists of a fluid *plasma* with red and white blood *corpuscles* and *platelets* suspended in it. Dissolved in the plasma are various inorganic salts, nutrients, hormones and excretory products. Other contents of the blood are mentioned later. The red blood corpuscles contain *haemoglobin,* which, when combined with oxygen, give the corpuscles the characteristic red colour. These cells have mostly lost their nuclei before they enter the blood circulation. White corpuscles do have nuclei and they are concerned with protection against infection — see p. 239.

Platelets are concerned with clotting of blood in wounds.

(e) *Nerve tissue.* Nerve cells, called *neurones,* have long projections, like arms, which have the special function of acting like roads for messages to pass along. Fig. 20.6 shows a neurone. The *dendrites* can pass messages (in the form of electrical impulses) towards the body of the neurone; the *axon* (usually only one per neurone) can conduct pulses only outwards from the body of the neurone.

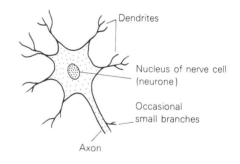

Fig. 20.6 A neurone

*Also, in smooth muscle each fibre is a single cell (but without a membrane). In striated muscle each fibre contains many nuclei and so is not a single cell and the fibre has a membrane around it.

The skin

A cross-section through the skin is shown in Fig. 20.7.

The *epidermis* is an epithelial tissue. New cells are being formed continually by mitotic division in the lower part of the epidermis (the germinative layer), and they push the cells above them outwards. The cells of the skin become flattened as they get nearer to the skin surface and lose their nuclei. Also, as they move towards the surface, they are gradually becoming filled with a special polypeptide called *keratin*. Soon the cells die and they are worn away from the skin surface. In this way skin is lost and is replaced from below. Complete replacement occurs over a period of something like 30 days.

The *dermis* is composed of areolar tissue with adipose tissue (subcutaneous fat) beneath and contains blood vessels and nerves.

The *hair follicle*, which is the tube that holds the hair, is really epidermal tissue pushed down into the dermis.

The muscle for erecting the hair is of little importance for the human but it can make hair 'stand on end' when one is cold or frightened.

Keratin is responsible for the horny nature of the cornified layer of the epidermis and improves the protection that the skin gives to the body. It also helps to keep down the loss of water from the skin so that it does not get too dry.

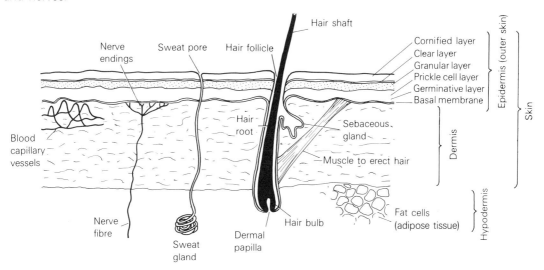

Fig. 20.7   The structure of the skin

The sweat glands

These are of two kinds, namely *eccrine* and *apocrine* sweat glands.

The eccrine type are found all over the body. They produce a clear fluid that is secreted by the cells of the long, tubular glands and passes out through pores on to the skin surface. This sweat is almost entirely water, but it contains some sodium chloride (common salt), and

some other chemicals such as *urea* which is a waste product of the body. The controlled release of sweat and the subsequent evaporation of its water gives the body an effective means of cooling itself whenever the surrounding temperature is high or the body needs to lose a lot of heat, for example, during exercise (Fig. 20.8, overleaf). Also sweating can occur when we feel nervous.

At other times there is little *sweating* (*sensible perspiration*), but evaporation of water from the moist skin, called *insensible perspiration,* still occurs and provides some cooling.

The second type, i.e. the apocrine sweat glands, are found, for example, in the armpits and on the sex organs. They are similar in structure to the eccrine type but larger and more coiled and they open into hair follicles instead of directly to the skin surface. The sweat from these glands is a fatty liquid often with a yellowish colour.

Unfortunately, bacterial decomposition of this type of sweat soon makes it become smelly and unpleasant.

Fig. 20.8   Cooling by eccrine sweat

The sebaceous glands

These glands open into the hair follicles and secrete an oily fluid called *sebum*. It keeps the skin soft and hair flexible. Insufficient sebum can lead to chapped skin.

The hair

As shown in Fig. 20.9, a normal hair on the head or elsewhere is contained in a tube-shaped pocket called a *follicle*. This follicle consists of an inner sheath and an outer sheath which are epithelial tissues like the epidermis of the skin. The outer sheath is a continuation of the germinative layer and is surrounded by dermal connective tissue. Lining the follicle is a scaly layer called the *cuticle*. The hair grows from the bottom of its follicle by mitotic division of germinative tissue that receives nourishment from the *dermal papilla* (Fig. 20.9).

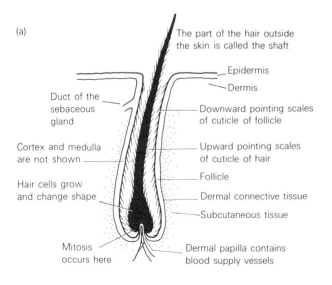

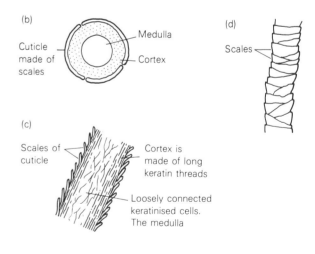

Fig. 20.9   Growth of a hair: (a) a section through a hair and its follicle;   (b) a section across a hair;   (c) a section through a hair showing its structure;   (d) the external surface of a hair very highly magnified

Growth of a normal hair

The life of a normal hair is divided into 3 stages which are known as *anagen, catagen* and *telogen.* The hair shown in Fig. 20.9(a) is at a late anagen stage. Soon after this the dermal papilla separates from the hair root, and so the hair is a *club* hair. Then, together with the surrounding inner sheath, it moves up towards the surface of the skin. The bottom of the follicle shrinks and some cells fill the space between the hair root and the dermal papilla. This is the catagen stage and usually lasts some weeks.

Following this there is usually a resting stage, the telogen stage. The hair root is not deeply embedded now, it is deprived of sustenance from the dermal papilla and it separates finally from the surrounding follicle. During this stage hairs are easily lost (mostly when the hair is combed or brushed), each hair lasting perhaps a few months if an eye lash or perhaps a few years if the hair is on the scalp. Below the club hair and the surrounding upper part of the follicle, there remains only the newly arrived dermal tissue and the dermal papilla.

When the new hair is to be formed, chemical stimulation occurs and the anagen stage begins by the mitotic division of the dermal tissue above the dermal papilla. The new cells produce a new lower end to the follicle and also a new hair that grows up through the follicle. As the cells of the hair move up the follicle they change shape to form the medulla, cortex and cuticle of the hair; they also become keratinised so that the hair acquires a reasonable stiffness, although it remains sufficiently flexible. Fig. 20.10 shows three stages in the life of a hair.

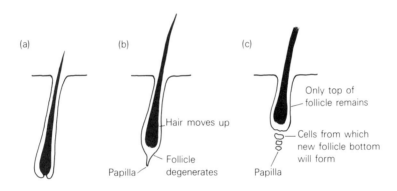

Fig. 20.10   The stages of hair growth: (a) anagen;   (b) catagen;   (c) telogen

Vellus hairs

Normal hairs are described as *terminal hairs. Vellus hairs* are different. They are small, soft, usually not pigmented, and grow only slowly. They are shallow-rooted, growing from a sebaceous gland, and receive poor nourishment. However, such a hair can under some circumstances accelerate in growth, deepen its root, acquire pigment, and become a terminal hair. Vellus hairs are sometimes described as *lanugo* but this term is best reserved for the fine hair seen on the body of a new-born baby.

Electrical epilation

Removal of hair by use of an electric current requires that the hair root is burnt by caustic soda produced by the galvanic method (the traditional electrolysis method), or by heating when the diathermy method (which uses a high-frequency current) is employed. For a permanent removal, the lower part of the follicle and the dermal tissue at the bottom of the hair root must be killed.

Some properties of hair

Keratin molecules in the hair consist of long spirals, like springs, which mostly lie parallel to each other with some connections between them, *cross-linkages,* caused by bonding between sulphur atoms in adjacent molecules. In this condition the keratin is called *alpha-keratin.*

Hair can be stretched when dry, but when wetted in cold water stretches more easily because the water molecules weaken the cross-linking bonds and make movement of the keratin molecules easier. If the hair is then dried while stretched, it will stay stretched without pulling for some time until the molecules gradually resume their normal lengths. Keratin with its molecules lengthened is called *beta-keratin.*

Hot water or cold alkaline solution produce a temporary set whereby the beta-keratin changes to alpha-keratin much more slowly. Using very hot water (almost boiling) or steam, cross-links are broken and new ones formed when the keratin is in the beta condition so that a permanent set is obtained. A further method of permanent waving uses a chemical (a *reducing agent*) to combine with the sulphur atoms that are responsible for the cross-linking; the cross-linkages are reformed after the hair has been suitably styled, by treating the hair with an *oxidising agent.*

Hair colour

The *colour* of the hair is due to the presence of granules of *melanin* in the cortex or, in the case of red or blonde hair, to *phaeomelanin.* Grey or white hair is caused by the absence of these pigments in the hairs.

Artificial colouring of hair is achieved by putting a coloured coating over the hair cuticle (surface dyeing) or by using a penetrating dye that enters the cortex and gives a more permanent and natural colour.

Skin colour

The skin usually shows a pinkness due to the red blood in the vessels of the dermis. This can become a redness (*erythema*) when the vessels dilate, as when a person is hot, or when there is inflammation of the tissues. Also, there is a brownness or even blackness of the skin which is caused by the presence of melanin.

Melanin is produced by melanosomes that are created by special cells called *melanocytes* in the skin's germinative layer. The formation of this pigment is usually the result of exposure to UVB radiation, often from the sun.

The melanin is made from a protein, *tyrosine,* with the help of an enzyme called *tyrosinase.* The part played by the UVB radiation is presumably to activate the tyrosinase and to cause the melanocytes to produce melanosomes. At first the melanosomes are colourless, but they darken due to formation of melanin and pass from the melanocytes into nearby cells of the prickle-cell layer (see Fig. 20.7). As a result, the skin shows its typical suntan colour. As explained in Chapter 17, this process called *delayed tanning* takes a few days to

develop fully, and soon the melanocytes become inactive again. The tan fades because the melanin-containing cells are lost since the whole skin is replaced within about 30 days. However, some tan remains after this period and some even remains as a permanent tan. The reason for this seems to be unknown.

We have discovered in Chapter 17 that UVA radiation in large doses can also cause delayed tan but, more interesting, it produces immediate tan. This tan is largely explained as the completion of the final stages of melanin production (and is limited by there being no creation of fresh melanosomes). The immediate tan is at its best within an hour following exposure. Ultra-violet irradiation does not affect the number of melanocytes present; but UVB does decide how many are active. The question of whether a person is fair-skinned or dark-skinned is decided by heredity and is related to the size of the melanin granules produced; a black person has much more melanin in the skin.

What use is sun-tan to the body?

Certainly melanin absorbs ultra-violet radiation and, together with the thickening of the horny layer of the skin as the result of exposure to UV, this protects the living cells of the skin from harm by subsequent doses of UV. Alternatively, it has been suggested, the melanin production and fading might have as its main function the control of vitamin D formation that is caused by UVB. Were it not for this a person could accumulate too much vitamin D.

Erythema

This is the reddening of the skin's appearance that accompanies the dilation of blood vessels in the dermis. It is commonly produced when a chemical called *histamine* is released from damaged cells, as when the skin is rubbed. Heating, including heating by infra-red radiation, also causes *erythema*. Erythema due to IR occurs within minutes, whereas erythema due to UV radiation can be delayed (see Chapter 17) and seems to be the result of chemical changes caused by the UV.

Nails

The *nails* of the fingers and toes are made of a special type of compact keratin and are really extensions of the clear layer of the skin (the *stratum lucidum*). The nail root grows from germinative tissue that forms the *matrix*. Much of the nail area appears pink because it is in good contact with the nail-bed, and light passes through easily so that the redness of blood in the dermis is seen. The keratin of the nails does not flake off but is removed by wear or by manicure. The structure of the nail is shown in Fig. 20.11.

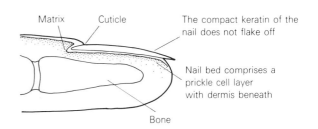

Fig. 20.11 The structure of a nail

MOVEMENT

Bones

Bone is classified as a connective tissue, but it contains a lot of mineral material in the form of carbonates and phosphates of calcium which make it hard and fairly rigid. The fibres that characterise connective tissues are not very noticeable in bone. Fig. 20.12 (overleaf) shows the construction of a long limb bone. The bony material is of two types, namely, *compact bone* and *spongy bone*.

The compact bone is perforated by narrow channels through which blood vessels and nerve fibres are able to pass and the mineralised tissue surrounds these channels in layers with bone cells lying between the layers.

The bone cells are of two kinds. One kind is responsible for producing the bony tissue, and the other kind breaks it down. It is by the correct balance of these two actions that bones can grow.

The spongy bone consists of thin pieces of mineralised bone with red marrow in between them. This marrow is important because it produces blood, as explained later in this Chapter.

The cartilage around the ends of the bone (commonly described as *gristle*) provides a smooth surface where the bone is jointed to the next bone and so friction is minimised.

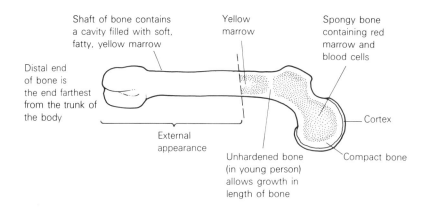

Fig. 20.12   The structure of a bone — the femur

The skeleton

Bones are linked together to form the framework which we call the *skeleton*. It serves to support the body, protect the heart, lungs and abdominal organs, brain, etc., and by virtue of the joints and muscles attached to the bones the skeleton allows the body to move around. The bones of the skeleton are seen in Fig. 20.13.

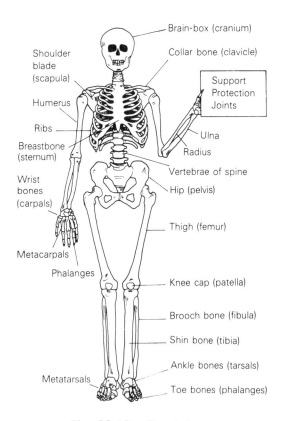

Fig. 20.13   The skeleton

Joints

Where two bones meet we have a *joint* and joints may be of different types according to the movement that the joint permits. There is: (a) the *hinge type* such as the knee or elbow, (b) the *ball-and-socket type* where the limbs meet the trunk, (c) the *pivot type* where one bone keeps still while the other pivots round it (the skull turns on the top of the spine), (d) the *gliding type* (the *spine*) where each of many bones moves a little to provide a total movement that is appreciable, (e) some joints permit little or no movement and the bones of the *brain-box* (cranium) are of this *immovable* type.

Cartilage

Cartilage is a firm but non-rigid connective tissue and occurs in three forms, namely, *hyaline* ('glass-like') *cartilage, elastic cartilage* and *fibrocartilage.* The hyaline cartilage is translucent and contains collagen fibres. It is found covering the ends of bones where they form joints (*articular cartilage*), as in the joints of the limbs, in the nose, voice box and windpipe. Elastic cartilage is yellow in colour due to the elastic fibres which it contains. It is found, for example, in the outer flaps of the ears.

Fibrocartilage is a denser tissue than hyaline cartilage, containing a greater concentration of collagen fibres. It is found between the vertebrae of the spine.

Cartilage is enclosed within a membrane which is supplied with blood vessels and it is from here that the cartilage obtains its nutriment.

Ligaments

These are connective tissues that contain mainly the yellow elastic fibres and occur at joints where they hold together the bones forming the joint, as shown in Fig. 20.14.

The capsule of *synovial fluid* enclosed within the synovial membrane shown in the diagram, acts as a lubricant at the joint so that the ends of the bones do not in fact touch each other.

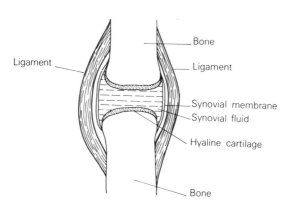

Fig. 20.14   Ligaments hold a joint together

Muscles and tendons

The function of *muscles* is to contract to produce movement (and some have to relax to allow other muscles to produce movement). The *tendons* are tissues which connect muscles to bones. They are connective tissues made up largely of inelastic fibres so that they do not stretch.

As shown in Fig. 20.15 (overleaf), the end of the muscle which is connected to the bone to be moved is called the *insertion*, while the other end of the muscle is the *origin*.

235

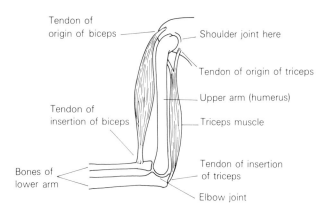

Fig. 20.15 The muscles of the upper arm

Muscles that bend a limb, such as the *biceps,* are called *flexors* and those that straighten the limb, such as the *triceps,* are called *extensors.*

A muscle causing a movement such as the biceps raising the lower arm is called a *prime mover,* while the muscle that must relax during this movement, the triceps in our example, is called the *antagonist.* In addition, other muscles may need to contract to prevent unwanted movements that might otherwise be caused by the contraction of the prime mover.

As mentioned earlier in the Chapter, muscles may be smooth or plain. Also, there is the type of muscle found in the heart which is rather special.

Muscles such as those of the limbs that contract when one chooses are called *voluntary* muscles and are plain. The *involuntary* muscles of the internal organs, like those of the stomach and intestines and some of the lung tubes (but not those of the heart) are smooth muscle which is particularly suited to slow but prolonged contraction.

Muscle tone

As long as a person is conscious the voluntary muscles are always slightly contracted, even when no movement is required. This condition is called *tone*. It is the muscle tone that maintains the posture of the body. Especially, we don't fall to the floor when standing still because of the tone of muscles, namely, the *antigravity* muscles. Muscle tone also influences the shape of the body.

Muscle contractions*

(Details of how muscles contract may be found in the footnote.*)

In a voluntary muscle, a stimulating rate above 30 Hz is fast enough for there to be no sign of relaxation between the twitches so that the muscle remains steadily contracted. This is a condition called *tetanus.*

A faradic current, as defined in Chapter 4, therefore produces tetanus. A surging faradic current overcomes the problem of acclimatisation, and a rest period between surges not only helps to reduce acclimatisation but also reduces fatigue. Below 30 Hz tetanus is incomplete, the contractions are rather weak and all the fibres do not contract at the same time; individual contractions can be observed. (This is true of interrupted galvanic currents as defined in Chapter 4.)

---

*Earlier in this Chapter it was said that striated muscle is made up of long, thin cells called fibres and containing many threads *(myofibrils)*.

It can be shown too that even these fine threads are made up of still finer threads *(mycellae)* of *actomyosin* (a combination of the proteins *actin* and *myosin*) and it is these that have the unique property of being able to contract.

When a voltage pulse above a certain size (the threshold value), and of sufficient duration, comes through a nerve to a striated muscle it triggers the contraction of the muscle. The contraction, once started by the impulse, grows to full strength within a few milliseconds (the *latent period*), lasts perhaps 40 ms (the *contraction period*) and then gradually slackens off taking a similar time (the *relaxation period*). If, however, further pulses are sent to the muscle while it is contracted, the muscle will stay contracted until the pulses stop or until it becomes accustomed to the pulses and responds less (called *acclimatisation*) or until it tires *(fatigue)*.

A stronger stimulus, caused by a larger number of impulses, coming through the nerve causes more fibres to contract so that the muscle pulls more strongly.

The contraction is particularly strong if each stimulating impulse arrives when the contraction or 'twitch' due to the preceding impulse is at its maximum. This requires a frequency of perhaps 30 to 60 Hz depending on the muscle concerned. Pulses produced normally by the body are in the range of 5 to 50 pulses per second.

Smooth muscle experiences complete tetanus at quite low frequencies.

This electrical stimulation of a muscle may be direct, meaning that the current passes through the muscle to cause contraction, or it is indirect and acts on a nerve so that impulses are sent to the muscle through the nerve.*

Muscle fatigue occurs if there is a build-up of lactic acid in the muscle. This acid can be oxidised to carbon dioxide and water or it is carried away by blood and then converted to glycogen, mostly in the liver.

Muscle tone is maintained by quite high-frequency impulses but fatigue is avoided because not all the fibres are contracting at any one time. Many are resting while others contract.

When a muscle is warmer it responds more quickly and more strongly. It also gives a stronger contraction if it is loaded — for instance, lifting a weight or pushing or pulling.

The efficiency with which muscles contract is usually about 20%, meaning that the energy (obtained from glucose phosphate which comes from glycogen) becomes partly the kinetic energy of movement while 80% of it becomes heat in the muscle. In the case of isometric contraction (see Chapter 1) all the energy becomes heat since no useful work is done.

Oxygen debt

During vigorous exercise, muscles produce lactic acid much faster than it can be removed because oxygen cannot be supplied fast enough. Although fatigue will set in, a considerable build-up of the acid can occur first.

The oxygen needed to deal with this will have to be supplied after completion of the exercise and is called *oxygen debt*.

THE BODY'S 'DELIVERY SERVICE': OXYGEN AND OTHER CHEMICALS

The lungs

Fig. 20.16 (overleaf) shows that the body possesses two lungs each of which is formed of a large number of small chambers called *alveoli*.

Air is breathed into the lungs, passing into the mouth or nose, into the windpipe, and then through the *bronchi* and smaller *bronchioles*. Breathing out allows air to escape by travelling back along the same route and leaving via the mouth or nose.

In the body the lungs contain air and are always stretched because the pressure in the surrounding pleural cavity is low.

During breathing in (*inspiration*), the muscular diaphragm pulls down and some outward movement of the ribs occurs so that the pleural cavity pressure falls still lower,

the lungs expand and air pushes in. During *expiration*, the diaphragm and ribs move back so that the lungs, which are quite elastic, are able to contract to their original size and, in doing so, force air out.

Continual breathing in and out (called *respiration*) ensures that the alveoli always contain some fresh air. Oxygen in this air dissolves in the moisture of the thin walls of the alveoli and diffuses through these walls into the blood that flows in very fine 'capillary' blood vessels that cover the outside of every alveolus. At the same time, carbon dioxide passes out from the blood into the air in the alveolus. Thus the expired air contains an increased amount of carbon dioxide, which the body does not want, and has a reduced oxygen concentration.

---

*The particular chemical reaction responsible for contraction is the combination of ATP with glucose (from glycogen) to form ADP and glucose phosphate. Reactions like this are rapid and result in the sudden contraction of the protein polymer molecules which are the mycellae in the muscle fibres.

This is followed by a slower reaction where ADP combines with phosphocreatin recreating ATP, the other product being creatin. These processes do not require oxygen (they are anaerobic).

Subsequently the creatin reacts with glucose phosphate recreating phosphocreatin and producing lactic acid. This last process requires oxygen.

The blood that came to the lungs with relatively little oxygen and too much carbon dioxide is now oxygenated; it is returned to the heart via the pulmonary veins ready to be sent to all parts of the body.

The breathing rate is typically about 20 per minute in adults. It is basically an involuntary mechanism requiring no thinking at all, but is subject to voluntary control to some extent, for example, when the breath is held or talking or singing occurs. Normally the breathing rate is largely controlled by the carbon dioxide concentration in the blood, a higher concentration causing faster breathing.

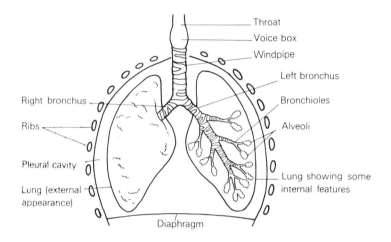

Fig. 20.16   The lungs

The blood

As explained in Chapter 2, blood flows from the heart to all parts of the body through *arteries* and returns through *veins*. This circulation of blood takes oxygen and other nutrients to all the body's tissues and removes carbon dioxide produced from all the energy-releasing oxidations. The water content of tissues can be kept correct by supply of water from the blood and heat is supplied or removed according to whether the tissues are cooler or warmer than the blood passing through.

Chemicals and water enter and leave the blood through the thin walls of the very slender and very numerous *capillary* vessels into which the arteries divide. It is the pumping of the heart, typically 70 times per minute, which sends blood through arteries. These branch into smaller vessels called *arterioles* and then into the very fine capillaries. The blood circulation is illustrated in Fig. 2.8 (Chapter 2) and in Fig. 20.17.

Arteries are strong, elastic, thick-walled tubes. The walls of the smaller arteries and the arterioles contain more muscle and this can contract to control the flow of blood as necessary. Once through the capillaries blood returns to the heart through veins. These have walls that are thinner than those of arteries. Some veins, such as the larger ones in the legs, have to transport blood upwards against gravity. They have valves at intervals along their lengths which allow blood to flow in the required direction only.

The *blood* is made up of the clear, slightly yellowish *plasma* which comprises water with some proteins, organic waste products (such as urea), organic nutrients (amino acids, glucose and fats), mineral salts, enzymes, hormones and dissolved gases (oxygen and carbon dioxide). Carried in the plasma are the *red corpuscles*, the *white corpuscles* and the blood *platelets*.

Red blood corpuscles (*erythrocytes*) are round, flattened cells that are also flexible. They contain *haemoglobin* which gives them their characteristic red colour. Haemoglobin is an iron-containing pigment which combines readily with oxygen if exposed to this gas and becomes *oxyhaemoglobin*. The oxygen is held only weakly, and the oxyhaemoglobin, unlike a true oxide, releases the oxygen again if it is exposed to a lower concentration of oxygen. Thus oxygen is taken up particularly by these red corpuscles when the blood passes through the lungs. In the body tissues plasma leaves the capillaries because the blood pressure is high and so conveys oxygen from the red corpuscles to the surrounding tissues. At the ends of the capillaries where they join to form veins the blood pressure is low so that plasma re-enters the blood due to osmotic pressure (see Chapter 9).

Carbon dioxide is taken away from the body tissues, partly combined with haemoglobin (forming *carboxy-haemoglobin*), partly dissolved in the plasma, and partly as dissolved sodium bicarbonate.

Red blood cells are formed in the red marrow of bones such as the ribs and the ends of long limb bones. They are then released into the blood circulation but, by this time, the nuclei have disappeared and the corpuscles gradually deteriorate in condition while they carry out their important work in the blood stream. After about 100 days a red corpuscle will break up. It is the spleen and liver that remove the fragments of the disintegrated red corpuscles from the blood stream.

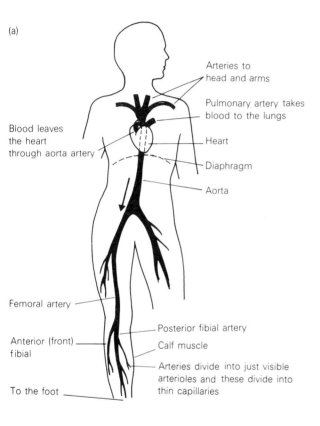

(a)

Arteries to head and arms

Pulmonary artery takes blood to the lungs

Blood leaves the heart through aorta artery

Heart

Diaphragm

Aorta

Femoral artery

Anterior (front) fibial

Posterior fibial artery

Calf muscle

Arteries divide into just visible arterioles and these divide into thin capillaries

To the foot

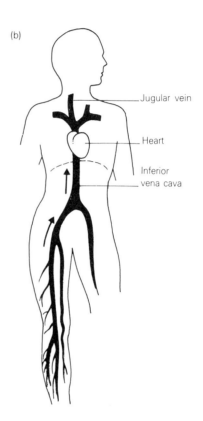

(b)

Jugular vein

Heart

Inferior vena cava

Fig. 20.17 The vessels of the blood circulation system: (a) how blood reaches the calf muscle; (b) return of blood from the calf muscle

White blood corpuscles

White corpuscles are fewer in number than red ones. They contain no haemoglobin and they all possess nuclei throughout their existence. They are not all of the same type, differing in their appearance and in their properties and being produced in different places which can be the red marrow in bones, the lymphatic system or the reticulo-endothelial system. They are important for

defence against any bacteria that enter the body, and their action is described in Chapter 22.

Blood platelets are smaller cells without nuclei, and are also formed in the red marrow of bones. They are concerned with clotting of the blood when a vessel is broken by a wound. This is also explained in Chapter 22.

Lymph

The liquid which passes out through the blood vessel walls into the surrounding tissues is called *lymph*. Cells of muscles, skin and other tissues receive their nourishment from the lymph. Waste chemicals and other chemicals from the cells pass into the lymph which is taken away through the lymphatic system (see Chapter 2).

Serum is blood from which corpuscles and fibrin have been removed. It is seen to ooze from wounds when clotting occurs.

Lymph differs from plasma in that it contains no proteins.

Lymph glands

Also called *lymph nodes*, these are seen as frequent swellings of the lymphatic vessels. These glands act as filters, preventing bacteria from passing through. In the nodes, white blood cells are produced which can attack germs in the lymph nodes but can also be carried by the lymph and then by the blood system to all parts of the body.

The reticulo-endothelial (RE) system or macrophage system

Not only are most white blood corpuscles able to eat bacteria (i.e. are *phagocytes*), but throughout most of the body's tissues other phagocytic cells are found that do not circulate in the blood. These cells can ingest not only bacteria but also dead cells, worn-out red blood corpuscles and comparatively large foreign bodies. They are called *macrophage cells*. Although found in most tissues to some extent, they are found particularly in connective tissue, lymph nodes, bone marrow, the spleen and the liver.

These macrophage cells are collectively called the *RE system* after the RE tissues in which they are mostly found, the reticular and endothelial tissues of the spleen, liver, etc.

The spleen is described below and the phagocytic action of white blood cells and macrophage cells is explained in Chapter 22.

The spleen

This organ is found close to the stomach. It has a bag-like appearance and, inside, its volume is divided into spaces called *sinuses* which are filled with blood and lined with macrophage cells.

The *spleen* acts as a store of blood that can be used to boost the rest of the blood system in an emergency. Also, its macrophage cells act upon any broken or damaged red blood corpuscles that come into the spleen and produce from these the pigments which the liver will subsequently take from the blood and put into the bile. Macrophages in the liver and red bone marrow can help in this process too.

Some white blood cells are produced in the spleen.

THE BODY'S 'TELEPHONE SYSTEM'

The nervous system

This is the body's means of fast communication, a sort of telephone system. A slower method involves the release of messenger chemicals (hormones) into the blood and this is discussed later in this Chapter.

The *nervous system* is made up of nerve cells (*neurones*) and it is along the long, thin extensions of these cells, i.e. along the nerve fibres, that messages travel throughout the body (Fig. 20.18).

The *brain* and the *spinal cord* are illustrated in Fig. 20.24. Together they are called the *central nervous system*. It is from this central nervous system that instructions are sent out. News of stimuli such as temperature changes, chemical changes, pressures, or what is seen, heard, tasted or felt is conveyed to the central nervous system in the form of voltage pulses, namely, *nerve impulses*, that pass along *afferent* nerve fibres. Nerve fibres that carry instructions from the central nervous system, to muscles, for example, are called *efferent* fibres.

Peripheral and autonomic nervous systems

Those parts of the nervous system outside the brain and spinal cord can be conveniently divided into two systems, the *peripheral system* and the *autonomic system*. The autonomic system is then divided up again into two parts, the *sympathetic system* and the *parasympathetic system*.

The peripheral nervous system

This system is distinguished from the autonomic system in that many of its messages result in conscious decisions being made (in the brain). Resulting instructions can be sent, as impulses, of course, to muscles that are required to contract.

As an example, consider when the right thumb is too hot or itches. The ends of nerve fibres, *nerve endings,* beneath the epidermis of the thumb are stimulated to produce voltage pulses. These impulses travel up the fibres which pass through the arm along with many other fibres forming the *median nerve*. The impulses then travel by one of the routes shown in Fig. 20.18 and move along one of the spinal nerves that pass between the bones of the spine and enter the spinal cord. From here the impulses pass up the spinal cord to the brain where thinking takes place, and perhaps it is decided that the thumb should move. Consequently, impulses are sent through the appropriate efferent fibres to the muscles that will move the thumb. They travel down the spinal cord, out through a spinal nerve, through the brachial plexus (a network of nerves), down the median nerve and so to the muscles at the base of the thumb.

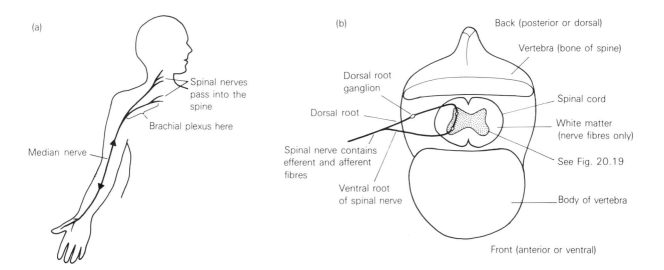

Fig. 20.18   The nerve supply to the thumb: (a) the median nerve;   (b) how the spinal nerves link with the spinal cord

241

The reflex arc

Sometimes a sudden stimulus produces an immediate muscle response, in fact a jerk, with no thinking involved. This is a *reflex action*. It occurs by an impulse travelling along a path called a *reflex arc*. An example would be when a person jumps up after sitting on a pin or blinks when something comes near to the eye.

In these examples the responses are made in the spinal cord. In many reflex arcs, impulses travel to the spinal cord by afferent neurones, through connective neurones in the spinal cord, and then through efferent neurons to the appropriate muscles (as suggested in Fig. 20.19).

Mouth-watering at the sight of food is a reflex action but the response is produced in the brain rather than the spinal cord.

However, most reflexes involve more than three neurones, and many afferent and efferent fibres will be used. Also, in most cases connections are present between the reflex arc and the brain so that the person feels the stimulus although the response has already been decided.

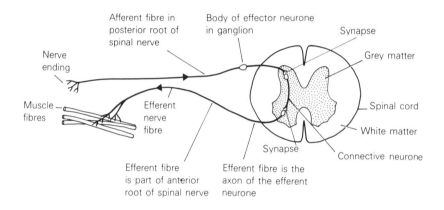

Fig. 20.19 A reflex arc

Synapses

When an impulse reaches the end of an axon of one neurone it can only move on through a dendrite of a further neurone by a chemical action occurring at the place, called a *synapse*, where the two fibres almost make contact. The chemical that is released when the impulse arrives at the synapse is acetylcholine in many nerves, but is noradrenalin in certain sympathetic nerve fibres.

Messages travelling through the nervous system move more slowly across synapses than along the fibres, so it is this movement that mostly determines the time taken for the body to react to a stimulus. The synapses are

important too because they allow impulses to move in one direction only, away from the central nervous system in efferent fibres.

Except for the existence of synapses, impulses can travel in either direction in any fibre. A consequence of this is that an impulse can move along a fibre towards the central nervous system but before reaching a synapse it can pass down a side branch and travel away from the central nervous system; then it can cause a response as if it had come from the brain or spinal cord. Stimulation of the skin can result in dilation of nearby blood vessels because of such a process.

Structure of a nerve fibre

Many nerve fibres are constructed as shown in Fig. 20.20. The *myelin sheath* is absent from some nerves

and also disappears at the ends of fibres when they reach, for example, the muscle which they control.

When a nerve fibre carries information towards the central nervous system the fibre is described as *afferent* or *sensory*. Obviously news of what is seen, heard or felt passes along sensory fibres.

Efferent fibres can carry impulses that make muscles contract or make glands secrete chemicals; they are *motor fibres*.*

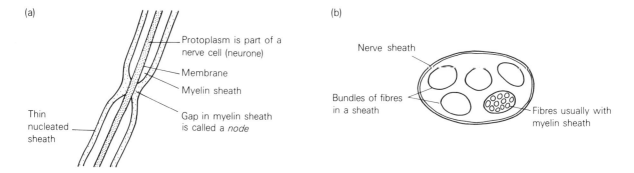

(a)

Protoplasm is part of a nerve cell (neurone)

Membrane

Myelin sheath

Gap in myelin sheath is called a *node*

Thin nucleated sheath

(b)

Nerve sheath

Bundles of fibres in a sheath

Fibres usually with myelin sheath

Fig. 20.20 The construction of nerves and nerve fibres: (a) part of an axon; (b) a nerve consists of many fibres

How a nerve impulse is produced

Normally a fibre is electrically negative inside while the outside of its membrane is positive. Both inside and outside the fibre there are ions, some of which can move freely through the cell membrane at nodes. The fibre settles down with its inside negative while outside its membrane the potential is positive. Sodium ions are in high concentration outside the fibre but not inside and are unable to penetrate the cell membrane.

When a stimulus excites a nerve fibre the membrane becomes porous to the + sodium ions and the ions pour in so that the inside of the fibre becomes less − or even +. This voltage change is the nerve impulse and it starts at a node because the myelin sheath does not let any ions pass through it. A strong, continued stimulus normally causes a series of impulses in rapid succession.

It is not surprising that excitation of a fibre at a node (or wherever there is no myelin sheath) can be achieved by the use of an electric current because this affects the potential difference across the fibre membrane.

Transmission of an impulse

The impulse makes current flow along the fibre and is enough to start off the same process, causing sodium ions to rush in, at the next node. Thus the impulse travels along the fibre. (See Fig. 20.21.)

The strength of this impulse does not fade along the length of the fibre because it is being renewed at every node. Neither is the size of the impulse affected by the strength of the stimulus as long as it is large enough and of sufficient duration to make the cell membrane porous.

The sodium ions that have entered the fibre will spread out but, to avoid a permanent build-up, these ions must be removed. This is achieved by a continual chemical action, in fact an oxidation, which requires energy. It is obtained by oxidation of glucose.

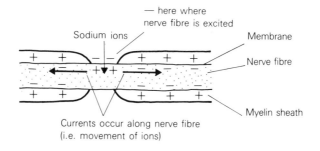

− here where nerve fibre is excited

Sodium ions

Membrane

Nerve fibre

Myelin sheath

Currents occur along nerve fibre (i.e. movement of ions)

Fig. 20.21 The production of a nerve impulse

Stimulation of muscle contractions

The voluntary, striated muscles in which a therapist is particularly interested can be made to contract by directly exciting them with a current passing through the muscle, or the electric current of suitable frequency can cause impulses to be generated in a nerve along which they then pass to the muscle. Here the nerve fibres divide up into many branches to form a 'motor end-plate' and each branch makes contact with a

Accommodation

If a steady current is applied in the hope of producing a continual succession of impulses, it soon loses its effect, as if the body were getting accustomed to it and taking no further notice of its presence. This phenomenon is called *accommodation*.[†]

For electrical stimulation an alternating current or pulses of current are used. In beauty therapy it is nearly always pulsed current that is used.

Choice of size and duration of stimulating current pulse

Each pulse must be of sufficient duration and strength to cause excitation and a greater strength requires less time. This is shown in Fig. 20.22.

The *rheobase* is the smallest current that will cause excitation. The *chronaxie* is the smallest time that can be used without using too large a current (usually defined as twice the rheobase). (Fig. 20.23.)

A faradic current of frequency 50 Hz maintained for a few seconds followed by a rest period of a few seconds

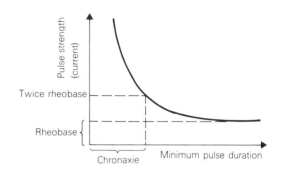

Fig. 20.23   Chronaxie and rheobase

and then repeated is typical for stimulating tetanic contraction of a muscle, with each pulse having a duration of 1 ms. The pulses are then following each other with 20 ms intervals between them which is sufficient for recovery after each nerve impulse. The duration of the pulses is equal to the approximate chronaxie (which is greater or less than 0.1 ms, depending upon the nerve fibres concerned).

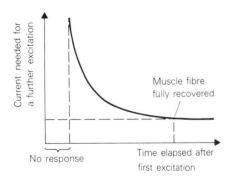

Fig. 20.22   The size of current needed for excitation of a muscle

---

muscle fibre. It is where the motor end-plate contacts the muscle that the impulses release a chemical (acetylcholine) which makes the covering membranes of the muscle fibres become permeable to sodium ions. Consequently a voltage pulse is produced, i.e. an impulse, and this spreads in the muscle just like a nerve impulse. It is the muscle impulse that initiates the contraction. This impulse is short-lived but the muscle contraction lasts for something like 100 ms because the muscle fibres continue to be excited by movement of ions within the fibre triggered by the muscle impulse. A muscle stays contracted only if impulses arrive in rapid succession.

Recovery after excitation of muscle

After producing one impulse, further excitation is impossible for a short time (perhaps 0.1 ms), followed by a time of about 10 ms during which a stronger stimulating current is needed to produce excitation. If, however, a time greater than this is allowed, the potential difference across the membrane will be returned to normal and excitation will occur easily again.

To keep a muscle contracted the series of impulses must have a frequency that allows recovery after each excitation.

[†] It can be explained by the ions in and around the fibre having time to settle so that the combined effect of the ions and the applied current is to produce the normal potential difference across the cell membrane.

Nerve stimulation only

When faradic currents have a frequency that is rather too high for muscles to contract, say 100 Hz or more, it is still possible for nerves to be excited; this can lead to dilation of blood capillaries and a general warming of the area.

Motor points

These are places where electrodes most easily produce excitation of a muscle or a nerve. A nerve *motor point* is usually where the nerve comes close to the body surface. A motor point of a muscle is usually an area close to where the motor nerve enters the muscle.

The autonomic nervous system

This part of the nervous system receives news and replies to it, but without one ever being aware of what is happening. It controls the rate at which the heart beats, the movements of the stomach, and the diameter of the small blood vessels. Breathing is another example: but, of course, breathing can also be controlled voluntarily and we can be conscious of the chest movements.

The two parts into which we divide the autonomic nervous system are the *sympathetic* and the *para-sympathetic nervous systems*. The first of these encourages the heart beat to speed up and the blood vessels in the muscles to widen so that the blood flow increases. These are the sort of changes needed in emergencies when a fight or flight might result. In

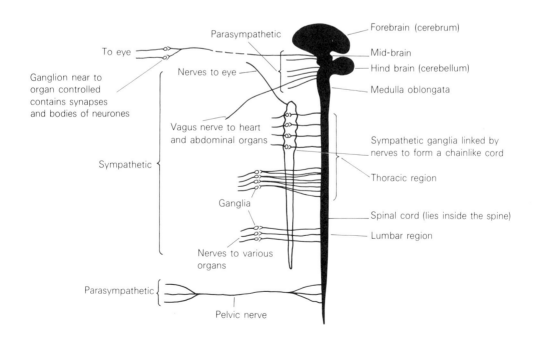

Fig. 20.24   The central nervous system and the autonomic nervous system

contrast to 'fight or flight', the parasympathetic system is associated with resting. It slows down the heart beat, and the blood supply to muscles is reduced.* The relation of the autonomic nervous system to the central nervous system is shown in Fig. 20.24.

An organ such as the heart, or a gland such as the pancreas, receives a nervous supply from each of the two autonomic systems, the one trying to make the organ more active, the other trying to do the opposite; the correct balance is hopefully maintained.

THE BODY'S 'MESSENGER SERVICE'

Endocrine glands

As explained earlier in this Chapter, the function of a gland is to secrete chemicals; examples of gland structure were shown in Fig. 20.4.

Glands are described as *exocrine* if they deliver their secretions to the outside of the body, or into the alimentary canal (which, in a way, is outside the body). Examples are sebaceous glands.

Other glands, called *endocrine*, liberate chemicals called *hormones* into the blood. Their locations may be seen in Fig. 20.25.

These hormones are best regarded as *messenger* chemicals because, as they travel to various organs of the body, they control the organ's activity, stimulating it or restraining it. This control is in addition to the control of organs by nerves. Instead of messages in the form of words the body communicates with chemicals, the hormones from the endocrine glands, and with nerve impulses (which we have seen are electrical) from the central nervous system.

The pituitary, the thyroid and the adrenal glands are important examples of endocrine glands.

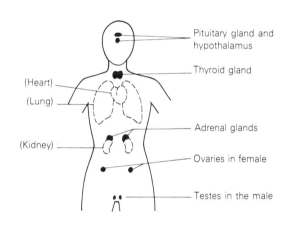

Fig. 20.25 The locations of endocrine glands

The thyroid gland

It is here that the hormone *thyroxine* is produced.

Thyroxine, a chemical containing iodine, spreads throughout the body via the blood stream and stimulates the metabolism. It speeds up the use of glucose. The extent to which thyroxine is released by the thyroid gland is controlled by a hormone secreted by the pituitary gland; the secretion of this thyroid-stimulating hormone is in turn decided by a part of the brain called the *hypothalamus*.

*The sympathetic and parasympathetic nervous systems also differ in another way. Some of the parasympathetic nerves come from parts of the brain while the pelvic nerve fibres come from the spinal cord, emerging between vertebrae in the sacral region of the spine. The efferent fibres of the sympathetic system all come from the spinal cord and pass out between the vertebrae of the thoracic and lumbar regions of the spine.

A further difference in structure between these two systems is that the parasympathetic nerves have swellings *(ganglia)* close to the organ which the nerve controls whereas the corresponding ganglia of the sympathetic system lie close to the spine.

The pituitary gland and the hypothalamus

The *pituitary gland* lies just below the brain, and the hypothalamus is a special part of the brain lying very close to the pituitary gland.

The pituitary is often described as the 'master gland' because it controls most of the working of other glands. The front part of this master gland secretes a hormone that controls the thyroid as already mentioned, a second one to control the adrenal glands, and a third that stimulates various activities of the reproductive system.

The pituitary also produces *growth hormone* which controls the growth of the child and, in the adult, seems to play some part in the control of the amount of glucose in the blood.

The hypothalamus controls the front part of the pituitary by itself secreting chemicals into the blood which pass into the nearby pituitary gland.

Other hormones are released from the posterior part of the pituitary gland.

In many ways the hypothalamus examines the body's conduction. It responds accordingly, for example, to the amount of glucose in the blood or to the blood's temperature, initiating the necessary controls on the metabolism by sending out nerve impulses to appropriate parts of the body or causing the release of pituitary hormones.

The adrenal (or suprarenal) glands

These are located on top of each kidney and each comprises an outer part called the *cortex* and an inside part called the *medulla.*

From the cortex are released various hormones, one of which influences the concentrations of mineral elements in the body, while another encourages conversion of protein to sugar.

The *adrenal medulla* is responsible for production of adrenalin and noradrenalin.

The release of adrenalin into the blood causes the metabolism to be stimulated, the heart to beat faster, and glycogen in the liver and muscles to be converted to blood glucose. These effects are designed to prepare the body for an emergency.

Noradrenalin is similar to adrenalin in its effects.

The pancreas

While the major part of the *pancreas* functions as an exocrine gland sending digestive juices into the intestine, it also has an endocrine part (groups of cells called the *islets of Langerhans*) that produces the hormones *insulin* and *glucagon.*

Insulin helps glucose to enter cells from the blood and enables glucose to be converted to glycogen. Glucagon causes release of glucose into the blood.

The sex glands

The *testes* and *ovaries* produce sperms and eggs as explained earlier, but they are also glands because they produce hormones.

The male hormones are responsible for the characteristic voice, physique and hair distribution of the man, while the ovarian hormones, which produce the feminine shape, are important for the menstrual cycle and during pregnancy.

SUMMARY

- The body is largely made up of cells, each having a nucleus and surrounded by cytoplasm.

- Some body cells can divide by mitosis, for example, the germinative layer of the epidermis produces new cells to replace skin being worn away.

- Meiosis is necessary to produce eggs and sperms.

- The body tissues are:

  (a) epithelial, such as the skin, linings of internal organs and glands

  (b) connective tissues containing elastic and collagen fibres and, in the case of adipose tissue, stored fat

  (c) muscle

  (d) blood

  (e) nerve tissue.

- Cellulite is a condition where there is a concentration of fluid in the connective tissue beneath the skin. It gives the skin a spongy and bumpy appearance. It is mostly found in obese people.

- The skin comprises epidermis, dermis and hair follicles; sweat glands pass through the skin.

- The outer dead layer of skin is composed of keratinised cells, and flakes off.

- Sweat is produced to cool the body.

- Sweat glands are of two kinds, eccrine and apocrine. Eccrine sweat is watery and apocrine sweat is thicker. Apocrine sweat is more readily attacked by skin bacteria and so produces body odour.

- The life of a hair is made up of the anagen, the catagen and the telogen (or resting) stage.

- Vellus hairs are soft, small, shallow rooted, poorly nourished and grow from a sebaceous gland.

- Epilation destroys the hair root and the lower end of the follicle.

- Erythema is the reddening of the skin due to increased blood supply. It is accompanied by increased blood circulation and lymph flow.

- Bones are made of compact bone and spongy bone inside.

- Bone cells are of two kinds, one building up bone and the other destroying it.

- Cartilage (gristle) protects the ends of the bones (hyaline cartilage) and is also found in the ear flaps, nose and windpipe.

- The skeleton supports the body, protects important organs and provides joints so that movement is possible.

- Ligaments hold bones together at joints.

- Tendons attach muscles to bones.

- Muscles can be extensors or flexors, prime movers or antagonists.

- Voluntary muscles are made of striated or plain muscle while involuntary muscles are smooth muscle, except for the heart muscles.

- Muscle tone is the tightness of voluntary muscles when they maintain the posture of the body.

- For muscle contraction to occur electric pulses called impulses must reach the muscle, usually by passing along a nerve.

- Impulses at more than 30 per second produce continuous contraction of a muscle; this condition is called tetanus.

- Muscle fatigue is due to accumulation of lactic acid in the muscle.

- A warm muscle responds more quickly and more strongly.

- Oxygen debt means using more oxygen during exercise than the body takes in.

- The lungs take in air from which oxygen passes into the blood (the red corpuscles) and carbon dioxide from the blood enters the air which is then expired.

- Blood consists of red corpuscles, white corpuscles and platelets carried in the plasma.

- Arteries carry blood away from the heart. All but the pulmonary arteries carry oxygenated blood. Veins take blood towards the heart and all but the pulmonary veins carry deoxygenated blood.

- White corpuscles are important in the defence against germs.

- The reticulo-endothelial system contains phagocytes.

- The spleen acts as a store of blood. White cells are produced there and phagocytes break down old and worn-out red blood cells.

- The nervous system comprises the central nervous system (brain and spinal cord), the peripheral system and the autonomic system.

- The autonomic system is divided into the sympathetic system (fight or flight activities) and the parasympathetic system (which encourages rest).

- Synapses occur between the ends of nerve fibres.

- Electric impulses pass along sensory nerve fibres to the central nervous system and instructions, also in the form of impulses, are sent back along motor fibres to muscles and organs.

- A reflex arc is a simple path followed by impulses, often to the spinal cord and out through motor fibres without the brain being involved. Decisions are made without thinking.

- Motor points are places where electrodes can be placed over a muscle or a nerve to easily excite a muscle.

- Endocrine glands secrete hormones (chemical messengers). Examples are the pituitary gland, thyroid, pancreas, adrenal glands, and the ovaries or testes.

- Adrenalin is a hormone secreted by the adrenal glands. It stimulates the metabolism.

- Insulin is secreted by part of the pancreas. It helps glucose to enter cells from the blood.

EXERCISE 20

1. Distinguish between *meiosis* and *mitosis.*

2. Draw a labelled diagram to show: (a) the layers of which the skin is made, (b) a hair follicle.

3. Describe the parts of which blood is made up.

4. List *four* functions of the blood. (CGLI)

5. What is meant by *cellulite*?

6. Name the stages of growth of a normal hair.

7. State the names of the two types of sweat glands. (CGLI)

8. What is *lanugo hair*?

9. What is *erythema*?

10. Explain what is meant by: (a) *muscle tone,* (b) *tetanus.*

11. (a) Name the principal waste product of muscle action.

(b) What is the result of an accumulation of this product?

12. Describe *two* differences between arteries and veins.

13. Distinguish between the *peripheral* and the *autonomic* nervous systems.

14. What are motor points?

15. State *two* functions of the skeleton. (CGLI)

16. Where in the human body is an example of: (a) an immovable joint, (b) a hinge joint? (CGLI)

17. What is meant by a hormone? (CGLI)

18. State the main functions of: (a) the spleen, (b) the lymph glands.

21

COSMETIC CHEMISTRY

COSMETICS

The wide variety of cosmetics

While medicines are used to improve one's health, *cosmetics* are chemicals used to improve one's appearance, to avoid unpleasant body odours and generally to make a person feel good.

A great variety of cosmetics can be obtained and most dressing tables, like the one illustrated in Fig. 21.1, display the cleanser, toner, moisturiser, founda-

tion make-up, blusher, mascara, eye liner and lipstick that are commonly used in the daily make-up routine. In addition to these we have nail varnishes and removers, talcum powders, cover-up applications, hair applications, deodorants, anti-perspirants, hand creams, perfumes and toilet waters.

Fig. 21.1   The daily make-up routine

Cosmetic ingredients

While most people seem to use cosmetics successfully with little or no knowledge of their ingredients, it is

more satisfying to know a little about the chemicals used. It is also helpful in choosing the most appropriate cosmetic, especially when faced with a bewildering range of manufacturers' advertisements some of which can be quite misleading. (Fig. 21.2.)

Fig. 21.2 Advertiser's optimism!

Skin cleansers

Removal of make-up, sebum and other secretions from the skin and any dead skin cells, talcum powder or household dust mixed in with the skin's oils and moisture, is not only important for hygiene but is necessary for the successful application of fresh make-up. Washing with soap and water is very effective but special *cleansers* are designed for the purpose. These often contain oils which mix readily with the oils to be removed and dissolve solidified sebum that can form in the openings of the sebaceous glands.

Cleansing creams frequently consist of oil in water emulsions. A simple one could contain just mineral oil, beeswax, borax and water. Perfume can be added to this.

Mineral oils are obtained from crude petroleum by distillation; the mineral oils mostly used for cosmetics are the technical white oils. These are less volatile than petrol, for example, and resemble lubricating oils. The oil selected for a cosmetic may be a light oil, i.e. not too viscous, but usually is liquid paraffin.

For a cleansing cream, the mineral oil serves as an effective solvent for the oils and greases to be removed from the skin and it is inexpensive and free from smell. It has the advantage over a vegetable oil that it does not deteriorate due to bacterial action or due to oxidation like the vegetable oils do. Examples of the latter are almond oil, olive oil and castor oil. Vegetable oils also tend to be more viscous and sticky. These oils feature more in cold creams.

A formula for a cold/cleansing cream might be:

Mineral oil	50%
Beeswax	15%
Borax	1%
Distilled water	34%

The ingredients are chosen to give a consistency to the final product that allows it to be spread easily. The beeswax and borax react chemically producing what is really a soap, and this acts as the emulsifying agent.

An oil in water emulsion is used because it is a more effective cleanser than an oil/wax mixture alone. A greater proportion of water makes a less effective cleanser but a less stiff cream is obtained.

When the cleanser is applied to the skin, water evaporates producing a pleasant cooling (hence the term *cold cream*).

Because mineral oil removes a considerable amount of the skin's natural oils (largely from sebum and needed for the skin to retain its moisture), it is usual to include ingredients such as spermaceti, cetyl alcohol, vegetable oils or lanolin to help reduce water loss from the skin (these ingredients are emollients).

Another possibility for a cleansing cream is a mixture of hydrocarbon oils and waxes. Not being an emulsion it is an *ointment* rather than a cream.

251

A liquefying cleanser that melts when applied to a warm skin can be obtained by the inclusion of a wax of appropriate melting point.

Cleansing lotions usually consist of an oil in water emulsion of appropriate viscosity or a solution of detergent (for example, triethanolamine lauryl sulphate).

The way in which a cleansing cream is removed is shown in Fig. 21.3.

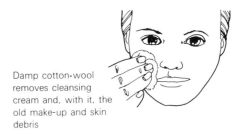

Damp cotton-wool removes cleansing cream and, with it, the old make-up and skin debris

Fig. 21.3   The removal of a cleansing cream

Toners

A *toner* is useful for removing the last traces of any grime, make-up or cleanser on the skin. The main ingredient of a toner is frequently an alcohol (such as *ethanol*) mixed with water, and the resulting toner acts also as an astringent and a cooling freshener. Other ingredients may have refreshing smells, for example, camphor, or, like the ethanol, be astringents, for example, witch-hazel. An antiseptic such as hexachlorophene may be put in and various perfumes are added.

For the dry or sensitive skin a lower concentration of alcohol is necessary and alcohol-free toners are also common to protect the skin.

Emollient creams

Emollient creams or lotions provide protection of the skin against loss of moisture. They do not put water into the skin but they reduce the loss of water (which enters from the underlying dermis). Emollients are particularly helpful when the air is dry and warm indoors, and in cold, windy weather. The skin is kept soft so that the term 'emollient' is appropriate. Almond oil and lanolin are common ingredients.

Moisturisers

These are usually creams or lotions applied after cleansing and toning and before foundation make-up. The *moisturiser* is an emollient, and it also provides a barrier between the skin and the make-up applied on top of it so that removal of the make-up later is easier.

The ingredient responsible for the emollient action of the moisturiser is usually *glycerine* which does take in some moisture from the air. Sorbitol is an alternative to glycerine. Both humectants because they retain water.

Vanishing cream

An emulsion of stearic acid in water can be used as an emollient and is described as a *vanishing cream*. For example, it could be made up of stearic acid, cetyl alcohol (a wax), an emulsifying agent (for example, triethanolamine will react with the stearic acid to form a soap), sodium hydroxide, distilled water, preservative and perfume. Preservatives are used to prevent decay due to bacterial action and fungus growth.

The emollient action is the result of the evaporation of the water leaving a thin, continuous layer of oil (stearic acid) covering the skin. This film reduces loss of water from the skin.

Vanishing creams may be used for foundation make-up, and hand creams, and as a base for other cosmetics.

Nourishing creams (skin foods)

These may be used with the intention of feeding the skin with ingredients such as oils, vitamins or proteins, but they are mainly emollients which work by leaving a thin film of oil on the skin surface. *Nourishing creams* are often applied as night creams.

Cetyl alcohol, spermaceti and lanolin are other useful emollients that may be included.

Foundation make-up

Foundation make-up is used to coat the skin so that powder can be applied and will stick. It is usually required also to give the skin an even colour, preferably the same colour as the skin or a little darker. A blusher or other colouring can be applied over this.

For a transparent foundation a vanishing cream can be used. A coloured foundation contains pigments such as zinc oxide, titanium dioxide or talc (which is a magnesium silicate), and these provide a white colour. Other pigments, iron oxide for example, are mixed in with these to create the required flesh colour.

These pigments can be held in a mixture of oils and waxes of suitable viscosity or can be mixed into an oil in water emulsion such as a stearic acid emulsion. Perfume is added.

Zinc oxide and titanium dioxide are renowned for their covering power and talc for its characteristic 'slip' which makes the skin feel smooth and the cosmetic spread easily. For a more translucent foundation a higher proportion of zinc oxide is used and less titanium dioxide. Carnauba wax can be added to stiffen the application to make a stick or cake.

Foundation make-up is also available in liquid and gel forms.

Sources of raw materials

The reader may already feel surprised at the number of chemicals used in cosmetics. More surprising perhaps are the places where they are found (Fig. 21.4).

Mineral oils, as has been said, are obtained from that dark petroleum oil taken from beneath deserts and seas. Talc can be mined (with some health risks). Lanolin is a grease taken from the wool of sheep and spermaceti is obtained from sperm whales. The sources of beeswax, almond oil and olive oil are obvious. Cetyl alcohol can be derived from spermaceti, and stearic acid from glyceryl stearate in beef fat.

Fig. 21.4 Some surprising sources of cosmetic materials

Face powders

These consist of finely ground, whitish powders mixed together with some coloured pigments added.

The user may require that the *powder* provides good opacity (covering power), that it hides or absorbs excessive sebum and sweat so preventing shininess, has a pleasant colour and, like other cosmetics, must stay on, produce no irritation, have no disagreeable smell and should apply easily and evenly. Translucency and sparkle are additional effects often required.

The formula below is a heavy powder suitable for an oily skin where shininess must be concealed:

Talc	55%
Kaolin	17%
Zinc oxide	15%
Zinc stearate	5%
Precipitated chalk	8%
Perfume and colour	

Talc has a rather low covering power but provides slip so that smoothness is obtained. Kaolin or fuller's earth (various silicates of aluminium) provides good covering power. It is particularly absorbent and reduces shine but, in large concentration, would be prone to streaking in damp weather.

Zinc oxide has very good covering power but the use of too much tends to be drying; titanium dioxide offers even better covering power, but does not stick on so well.

Precipitated chalk (calcium carbonate) has properties similar to kaolin but it also creates a pleasant bloom. Unfortunately it is characterised by its dry, chalky feel.

Zinc stearate provides good adhesion, is waterproofing and soothing, but too much can cause smears.

For sparkle, powdered metals can be added.

For compact form powders, as opposed to loose powder, a binder (which can be a gum) is mixed into the ingredients. The resulting powder is shaped into a cake by use of a press.

It should be pointed out that one should avoid breathing powders into the lungs. Fine particles of talc if inhaled in sufficient quantities over a long period are known to cause lung damage. Do not use powders carelessly (Fig. 21.5). Powder in the eyes can cause soreness.

Fig. 21.5   Do not handle face powder or talcum powder carelessly!

Blushers and rouges

Usually a *blusher* consists of pigments like talc or kaolin or titanium dioxide (let's call it a white pigment), mixed into a blend of oils and waxes; coloured pigments are mixed into this to give the required colour, red in the case of *rouge*.

The consistency of the blusher may be a cream, a semi-solid or a stick.

Dry rouge, made of powdered pigments mixed with a binder and pressed to form a compact, is an older formulation that can be used on a powdered foundation or the dried skin. It is applied with fingers or a puff.

Lipstick

Again a blend of oils and waxes has to be used as the base of the cosmetic. No single material offers the necessary ease of application and the required texture after application.

Mineral oils are poor solvents for colour, and some vegetable oils will undergo bacterial decay. Castor oil is suitable and has a high viscosity (avoiding smears and running of the *lipstick*), but it makes the lipstick drag when being applied. Many possible ingredients for lipstick are ruled out by their unpleasant tastes. Others are possible health risks, e.g. some organic pigments.

A simple formulation for a lipstick base is as follows:

Carnauba wax	12%
Beeswax	16%
Lanolin	6%
Cetyl alcohol	6%
Castor oil	60%

The proportion of waxes is high enough to produce the rigidity needed for a stick application. Cetyl alcohol and lanolin have emollient properties.

The colour in a lipstick can be provided by a suitable, coloured (usually red) pigment or a suitable stain that is water soluble so that it can enter the skin. Usually both are used together. An example of the first of these is carmine, obtained from the cochineal insect, but better, more opaque, synthetic pigments are now available. An example of the staining colour is acid eosin (orange) which dissolves in castor oil and changes to its red water-soluble form when it meets the skin (higher pH).

Perfumes are mostly avoided in lipsticks because the skin is easily irritated.

An antioxidant is usually included to prevent oxidation of the oils and a preservative to limit bacterial decay.

Eyebrow pencils

These are like lipsticks but with more wax and, of course, with different pigments.

Mascara

Once again, oils and pigments are mixed. Carbon black (really soot) can be used. Tiny fibres may also be included in the *mascara* and, when painted on to the eye lashes, the fibres add to the thickness and to the length of the lashes.

Eye-shadow

This can be a vanishing cream (an oil in water emulsion) with pigments added, or an ointment base can be used. Also, suitable powder applications are available.

Hand creams

Chapped hands are caused by loss of moisture from the skin, particularly in cold, windy weather because the relative humidity of the air is low. A water-containing cream applied to the skin is an obvious suggestion for combating this problem, but water very soon evaporates. Repeated immersion of the hands in water surprisingly can exacerbate the problem because soaking the skin makes it swell so that minute cracks probably develop in the horny layer. These cracks will subsequently allow escape of water vapour from the skin. Detergents, by removing skin oils, can also make the skin prone to drying.

Hand creams contain emollients and, as in moisturisers used on the face, they serve to reduce water loss from the skin and so allow the skin to recover its soft, moist texture.

Lanolin is a favourite ingredient in hand creams, but it tends to make the cream sticky if too much of it is used in the cream. Special care is needed in formulating a cream with lanolin in it because it can be a skin irritant.

The emollients used are usually made into a vanishing cream with a high proportion of water and a humectant included such as glycerol.

Face packs and masks

A *face pack* is an application such as the traditional mud pack that is spread over the face, left there for a while and then removed, usually by being peeled off. It is largely for the purpose of cleaning the skin.

A formula for such a pack is as follows:

Bentonite (aluminium silicates)	15%
Titanium dioxide	2%
Sulphonated vegetable oil	3%
Glycerol	5%
Water	75%

This pack or mask is applied as a paste 1 or 2 mm thick, and, as the water evaporates, the mask hardens and shrinks. It is this shrinkage that provides the desirable impression of the skin becoming tightened. The clay content (bentonite) absorbs the excess skin secretions (sebum, skin oils), while dead skin cells and other dirt (moist or oily debris) become embedded in the mask. The glycerol (a humectant) helps to retain some water and this, together with the sulphonated (emulsified) vegetable oil, keeps the mask flexible enough to be removed after re-wetting it.

Fig. 21.6   The use of a beauty mask

Masks may also include various additions such as astringent chemicals (for example, witch-hazel) and ingredients that are hopefully regarded as skin foods. It should be noted, however, that the skin cannot be nourished in this way. It can only be fed from the underlying dermis.

As an alternative to the paste mask, *liquid beauty masks* may be used. Typically such a mask contains a colloid that dries and shrinks to form a thin plastic skin. This pack provides mechanical astringency (due to its shrinkage) and cleanses the skin when the mask is peeled off, but very little cleansing by absorption occurs. Gelatin masks are of this type.

Some masks, usually called *cleansing masks,* do not dry and shrink and can be removed by wiping or washing off. Their action depends partly on the inclusion of slightly abrasive ingredients, and the paste is massaged into the skin sufficiently to loosen contamination from the skin. Ground almonds and various grain meals are common ingredients in these masks.

Fig. 21.7   An oil mask

Oil masks comprising a cloth soaked in warm oil can be applied to the face and the term *mask* is particularly apt in this case (Fig. 21.7). However, the term is usually regarded as synonymous with face pack.

Latex (rubber) masks use emulsions of latex in water which dry on the skin to leave a skin of latex which is subsequently peeled off. These and similar masks are impervious to water and are poor thermal conductors so that perspiration and erythema are encouraged and the skin fills out due to its increased moisture content. Here again very little cleansing by absorption occurs. The mask will, of course, remove fine hairs unless a suitable lubricant is put on the skin before the latex to prevent hair becoming embedded in the mask.

When a *wax mask* is used, the wax is melted and is then spread on to the face. It hardens as it cools and it acts like a latex mask, but with the added benefit of the heat from the wax.

Exfoliation

The use of a face mask removes some of the flaky surface of the skin, and so it can be described as *exfoliation.* Two other ways of producing exfoliation to obtain a smooth, fresh skin surface are either brushing with a suitable soap on the brush or massaging with an abrasive. The latter technique is known as a *facial scrub* and the abrasive can be simply an oatmeal and water paste. Alternatively, abrasive powders can be used in a cream or soap.

Depilation

Removing hair without destroying the lower part of the hair follicle is called *depilation*. It can be carried out by use of a chemical depilatory or by waxing. It is not a method of permanent hair removal.

Chemical depilatories work by breaking down the keratin of the hair.

The chemical chosen attacks sulphur bonds which link the long polymer molecules of the keratin. The same effect can occur in the skin but the effect there is much less, due to the lower concentration of sulphur bonds.

Thioglycollate salts, such as calcium thioglycollate, are commonly used. Possible ill-effects associated with calcium thioglycollate are bleeding under the skin, effects on the thyroid gland, and gastro-intestinal (alimentary canal) problems.

Depilation by waxing occurs when wax is applied like a face mask and the hairs become embedded in it. Removal of the hardened wax takes the hairs with it.

Leg waxing

Paraffin wax treatments can be used on the legs and arms to provide warmth and so give relief to aching joints or, alternatively, the aim may be to stimulate circulation and erythema as with a face mask. *Parafango wax* may also be used (a volcanic mud mixed with

paraffin wax), being preferred for its effective cleansing action. In this treatment it is usual to wrap the waxed limb in foil and then put towelling around this to stop rapid heat loss to the air (see Fig. 21.8).

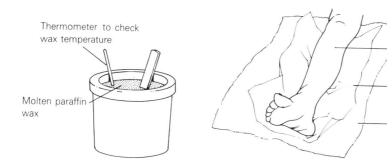

Thermometer to check wax temperature

Molten paraffin wax

Cream applied before waxing

Metal foil to wrap over wax

Towelling to wrap over metal foil

Fig. 21.8   Paraffin wax treatment

Nail varnish

What is required here is a liquid (or lacquer) that is painted on to the nails, dries quickly and leaves a hard, adherent, non-chipping film (the *varnish* or *enamel*). Usually a high gloss is required and colouring may be added.

Of course, nail varnish should not wash off nor should it harm the skin, and it should not stain the nails or skin.

Nitrocellulose provides a hard, tough film but its adhesion and gloss are not good, so resin is usually included to improve these two properties. Another additive, called a *plasticiser*, makes the film more flexible. The nitrocellulose is dissolved in a mixture of solvents which usually includes acetates such as butyl, ethyl or amyl acetate.

Adverse effects of nail varnishes include dryness of the skin and irritation around the nails, blackening and splitting of nails, skin rashes (for example, on eyelids), and nausea (feeling sick). The vapours should not be inhaled. All organic solvents should, if in doubt, be regarded as highly flammable.

Nail varnish removers

A solvent is used that dissolves the varnish, usually a mixture of solvents including butyl acetate and acetone. Careful use is required as for nail varnish because of the chemical and fire hazards (Fig. 21.9).

Fig. 21.9   Applying nail varnish — with care

Cuticle removers

These applications do not remove the *cuticle* from the nail, but they make it easy for it to be removed.

The best-known examples of *cuticle-remover* chemicals are weak solutions of sodium hydroxide or potassium hydroxide.*

A simple formulation could consist of potassium hydroxide dissolved in water, together with some glycerol to make the fluid more viscous, to reduce water loss and to minimise skin irritation that can result from the alkali. Perfume can be added.†

Sun-tan preparations

Various skin applications can be used with the aim of stopping erythema so as to prevent sun-burn yet allow a tan to be achieved.

Many organic chemicals are useful absorbers of ultra-violet. Commonly used for *sun-screen* preparations are esters such as the glyceryl ester of PABA (para amino benzoic acid). Sun-screens of this kind absorb a lot of the UVB radition while transmitting (allowing through) the small amount of UVA in the sunlight. As pointed out in Chapter 17, UVA can produce immediate sun-tan and some delayed tan with very little erythema.

Ideally, a sun-screen should not wash off too easily (because sun-bathers like to swim) and it should be non-irritant, without an unpleasant smell, not able to stain one's clothes and easy to apply evenly. The sun-screen mentioned above does not satisfy all these requirements.

The active agent of the sun-screen is put into a vanishing cream or other emulsion or into an oil so that an emollient action is also provided. Otherwise a solution of the sun-screen, for example, in alcohol, is used with an emollient to form a lotion.

An alternative kind of sun-tan application is the *total-block* sun-screen which uses a powder such as zinc oxide, kaolin or similar pigment in a suitable cream. The powder reflects all wavelengths of the sunlight. A thick

---

*These chemicals break down the keratin of the horny layer of the skin which forms the cuticle over the nail, achieving this by reacting with the sulphur bonds of the keratin.

†Cuticle-removing creams are common. The active ingredients may, for example, be included in a vanishing cream.

layer of this could stop all radiation reaching the skin but the thin layers usually applied reduce the amount of UV entering the skin (Fig. 21.10).

Fig. 21.10   A coating of a suitable sun-screen can stop sunburn yet allow UVA through

---
Calamine lotion

This lotion with its well-known pink colour is commonly used to soothe the skin if it has become sore because of sun-burn for example.

The ingredient which gives the lotion its name is zinc carbonate (*calamine*), but zinc oxide is mostly used instead. Some iron oxide is included and this provides the colour.

PERFUMES

How perfumes work

A *perfume* is a chemical from which vapour reaches the nose, dissolves in the moisture covering the lining of the nose and affects nerve endings there, creating a pleasing effect.

Sources of perfume

Traditionally, perfumes have largely been obtained from flowers and plants, and a few from animals.

These perfumes are mostly complex mixtures of odorous chemicals but individual components can be isolated chemically.

Other perfumes are entirely synthetic, not involving flowers, plants or animals at all.

Essential oils

These are naturally occurring, volatile substances with an appreciable smell, mostly obtained from plants.

Some well-known examples of these *perfume oils* are rose (obtained from the flowers' petals), bay (from the leaves), sandalwood and cedarwood (from the wood), cinnamon (using the bark), and lemon (from the peel).*

Many of them have antiseptic or germicidal properties.

Fixed oils

These are less volatile oils, such as castor oil, that are obtained from plants. They are mostly triglyceride oils.

Perfumes from animals

The familiar examples are *musk* which is obtained from the sex organs of the male musk deer, and *civet* which comes from the civet cat (Fig. 21.11). Other sources of perfume are the beaver and the whale. Some of us, perhaps because of our consciences, would prefer not to know this!

Fig. 21.11 Perfumes from strange places

Synthetic perfumes

Among these there is phenyl ethyl alcohol which is a rose perfume; benzyl acetate is a floral perfume, and various esters have a fruity smell.

Perfumes as applied to the skin

The term *perfume* is commonly used in two ways. Firstly, it can denote the liquid that is applied to the skin simply for the sake of its smell. This is usually a solution of perfume oil in alcohol. Secondly, the term 'perfume' can mean the perfume oil or essence as obtained from the plant or other source, and ready for dissolving in alcohol to make the skin application or for inclusion in various cosmetics.

Notes

The volatile components in a perfume cannot all be expected to be equally volatile so that, after application, the most volatile will evaporate more and will largely decide the odour of the perfume. This component is called the *top note*. The second, third, and so on, are the less volatile notes. For most of the perfume's life the fragrance is determined by the main note which is the 'body' of the perfume.

*They are soluble in alcohol but not very soluble in water and are mixtures of esters, aldehydes, alcohols, ketones and particularly terpenes

Fixatives

These are chemicals that may also be perfumes (like musk), which are used in a perfume to restrain the evaporation of one or more of the component perfumes. In fact, musk is usually used for its *fixative* properties. The aim is to make the components all evaporate equally so that the fragrance remains almost unchanged when it is used.

Extraction of essential oils

The processes used are distillation, solvent extraction, expression, enfleurage and maceration.

Distillation. A simple apparatus for *distilling* perfume oil is shown in Fig. 21.12. The essential oils evaporate and are collected first.

Steam distillation. This is the most common extraction method. Steam is bubbled through the mixture of water and plant material.

The steam enters the cells of the flowers, or other parts of the plants used, and vaporises the essential oil. The steam and oil vapour are then condensed and the oil is allowed to separate from the water.

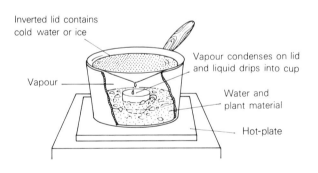

Inverted lid contains cold water or ice

Vapour condenses on lid and liquid drips into cup

Vapour

Water and plant material

Hot-plate

Fig. 21.12 Just about the simplest perfume still!

Fractional distillation. This technique has the advantage that the various oils from the plant material can be collected separately, the more volatile being separated from the less volatile.

Using a solvent such as ether. In this method the plant is soaked in a solvent such as ether (another very flammable organic solvent) into which the essential oils dissolve as well as some waxes and other chemicals. The ether is then evaporated, by a safe method, and the essential oil is dissolved from the distillate residue (what is left after the ether has gone) into alcohol.

Expression. This is simply squeezing the plant material to release the essential oils. The method is suitable for the skins of lemons, oranges and bergamots. (These are all citrus fruits.)

Enfleurage. What a nice word this is! It describes a method of extracting essential oils that is suitable for certain flower petals. Fat is repeatedly brought into contact with batches of fresh petals, and the perfume oil that has entered the fat is then dissolved out of it using alcohol.

Maceration. This is a grinding or stirring together of the perfume-containing material and a hot liquid, usually an oil.

Floral oils

n spite of the meaning attached so far to the term 'essential oils', it is apparently considered more correct, by some people, to reserve this term for those oils that have been obtained by distillation or expression and to describe the products of solvent extraction, maceration or enfleurage as *floral oils* (even if, sometimes, the essence is not derived from flowers).

Toilet spirit

This is the ethyl alcohol into which the perfume essence is dissolved.

It is highly *rectified*, i.e. well purified, containing only about 5% of water.

Ingredients of a toilet water

A *toilet water* is a solution of perfume essence in alcohol but with more alcohol in proportion to perfume, i.e. a weaker solution of perfume. Toilet water has a weaker fragrance which does not last so long as the normal perfume, and it is used more liberally. The alcohol used usually contains a high proportion of water, say 20%.

Maturing of a perfume

A period, perhaps of several weeks, is allowed for a perfume to age and *mature*. So it is stored in glass bottles with air and light excluded and then, after being cooled in a refrigerator, it is filtered to remove wax and other particles that have precipitated.

Massage creams

An emollient cream or lotion is commonly used for *massage* so that the fingers slide smoothly over the skin.

A cleansing cream may be used before massage is begun and, afterwards a toner/freshener may be used to remove the last traces of the massage cream.

SUMMARY

- Soap and water is an excellent skin cleanser.

- A skin cleanser can be a mineral oil in water emulsion plus an emollient and perfume.

- A toner removes the last traces of grime, make-up and cleanser. Its main ingredient is often alcohol. It therefore has an astringent and freshening action.

- A vanishing cream is a stearic acid in water emulsion.

- A moisturiser cream or lotion prevents loss of water from the skin and acts as a barrier between the skin and the foundation make-up. The emollient action is largely due to glycerine.

- Nourishing creams are essentially emollients.

- Make-up foundation is used to give colour to the whole face and powder must stick to it. A vanishing cream is suitable with pigments added to give the required colour.

- Zinc oxide is a white pigment, titanium dioxide is also white and more opaque, talc is white and provides slip, kaolin absorbs skin secretions, and calcium carbonate (chalk) gives bloom to a face powder.

- Lanolin and spermaceti are useful emollients and are obtained respectively from sheep wool and from whales.

262

- Face powders consist of finely ground white pigment such as talc, zinc oxide, titanium dioxide or kaolin.

- Face powders should not be breathed in or allowed in the eyes.

- Blushers and rouges usually comprise a mixture of white pigment with coloured pigments, mixed into a mixture of oils and waxes.

- Lipsticks are made of a mixture of oils and waxes containing coloured (red) pigment and dye to stain the lips red.

- Depilation is the removal of hair without much effect upon the follicle. It is not a permanent treatment.

- Depilation may be by peeling off a suitable wax application or by chemical depilation (using calcium thioglycollate, for example).

- Leg waxing is used as a heat treatment.

- A typical nail varnish consists of nitrocellulose as a film former, plus a resin to help adhesion, plus a plasticiser in a solvent such as butyl, ethyl or amyl acetate.

- Nail polish remover is a mixture of solvents such as butyl acetate and acetone.

- Many chemicals used for cosmetic purposes are highly flammable, poisonous or are skin irritants.

- Most sun-screen applications stop UVB but allow UVA through *or* they use pigments to reflect all UV. The latter is a total block sun-screen but is rarely thick enough to stop all UV.

- Natural perfumes are obtainable from flowers, plants and animals, for example, rose from flowers, cedarwood from trees, and musk and civet from animals.

- Essential oils are soluble in alcohol but are not very soluble in water. Many have antiseptic and germicidal properties.

- Perfume as applied to the skin is perfume oil dissolved in alcohol and mixed with water.

- Essential oils can be obtained from flowers and plants by distillation, by fractional distillation, by solvent extraction, by expression, by enfleurage and by maceration.

EXERCISE 21

1. Name an important ingredient of each of the following: (a) a skin cleanser, (b) a toner, (c) a moisturiser, (d) a foundation cream, (e) a face powder.

2. What is the source of lanolin? (CGLI)

3. What is a mineral oil?

4. Name *two* ingredients of lipsticks.

5. Distinguish between *epilation* and *depilation*.

6. Name *two* components of a nail varnish and explain the function of each.

7. On what principles are sun-screen applications based?

8. Name three ways in which essential oils can be separated from plant material.

9. Describe briefly a chemical reaction that produces a soap.

10. Give an example of a solvent used in a cosmetic preparation. (CGLI)

11. (a) State two differences between essential oils and fixed oils.

(b) Name one example of each type of oil.

12. Give two advantages of using liquid paraffin (mineral oil) in preference to a vegetable oil in a cosmetic cream.

22

HEALTH AND HYGIENE

HEALTH

Enemies of health

To the beauty therapist *health* must be thought of in its widest terms so as to include not just being free from sickness, injury or pain but also feeling good and looking good. The condition of a client's skin is of particular interest.

Health can be adversely affected by injuries, by faulty body metabolism, by infections of the body with bacteria, viruses, fungi or by parasitic animal life and even by environmental factors such as coldness or dryness of the surrounding air. Age, we know, also affects the condition of the skin and other tissues of the body.

Types of lesion

A wound or localised area of injury or tissue damage is called a *lesion*. Examples of skin lesions are cuts, scratches, abrasions (grazes), bruises, abscesses and warts.

A deep wound can cause dangerous loss of blood, serious or fatal injury to internal organs, and injuries to nerves or muscles, as well as infection by germs.

With shallow wounds our concern is the danger of infection and the possibility of a scar being left after the wound is healed. Scarring is also possible if epilation causes escape of serum and hence scab formation, and if the scabs are picked.

A *bruise* is a bleeding beneath the skin. A *blister* is an accumulation of liquid, largely serum, usually under the skin's horny layer. Such blisters are familiar as the result of friction. They can also be caused by burns and scalds. Small blisters are called *vesicles* and an example of these is cold sores on the lips.

Inflammation

This is a response to injury characterised by redness, swelling, warmth and often pain. Localised *inflammation* of the skin can result from rubbing, burns, scalds, insect bites, irritant or corrosive chemicals, or infection. It is a consequence of increased blood supply to the area of skin and an increased permeability of the blood vessel walls so that plasma escapes into the surrounding tissue.

The cause of this is, in part, the release of chemicals in the injured area. For example, *histamine* is released from some damaged cells, and this chemical causes the changes observed. Pain that may occur is due to stimulation of sensory nerve endings, partly because of pressure associated with swelling.

264

Redness, swelling and warmth that is superficial, particularly if it is not due to obvious accident or injury, is called *erythema*.

Abscesses

When a tissue is infected with micro-organisms (germs) such as bacteria, the intruding bacteria cause white blood corpuscles to rush into the area and start battle. Bacteria and some white corpuscles will be killed and some tissue cells broken down; it is these, together with plasma oozing from blood vessels in the area, that constitute the creamy fluid called *pus*. A lesion where pus is produced and collects is said to be *septic* and is called an *abscess*. A *boil* is an example of one (Fig. 22.1).

Acne is an affliction of the skin in which pus-containing pimples (*pustules*) often occur; boils and carbuncles

are larger abcesses which originate in hair follicles but extend into surrounding tissue.

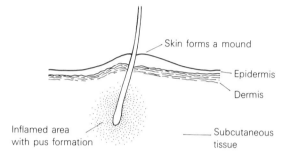

Fig. 22.1   A boil is an example of an abscess

Clotting of blood

When blood escapes from a wound, *coagulation*, or *clotting*, of the blood begins and, hopefully, this will soon stop the bleeding.

The clotting depends upon the presence of blood *platelets* and *fibrinogen* in the blood and it begins when platelets are damaged as the blood escapes.

The platelets are corpuscles, smaller than red blood corpuscles, and breaking them causes a chemical to be released into the blood. This chemical converts fibrinogen dissolved in the blood plasma into insoluble *fibrin* which

appears as fine threads radiating out from the platelets. The mesh of fibrin threads and platelets prevents the further escape of blood corpuscles, but *serum* (plasma minus fibrinogen) can ooze out. A jelly-like mass forms. This clot hardens to form a scab. Not only does this stop bleeding but it protects the underlying tissue and prevents entry of germs.

Various applications are sometimes used to speed up the clotting process and these are called *styptics* or *haemostatics*.

Healing of a wound

After a clot has formed it shrinks gradually, so pulling together the edges of the *wound*.

If the edges of the wound are in contact and no underlying tissue is missing, *healing* occurs without scarring. The wound heals by outward growth of the epidermis from its germinative layer and by growth from the sides

of the wound. This is true also of an extensive but shallow wound, like a graze. Where dermal and subcutaneous tissue is missing, fibrous connective tissue, blood capillaries and nerve endings replace it, but scarring can occur and the scar may be permanent.

Skin disorders due to imperfect metabolism

Common afflictions in this category are seborrhoea, blackheads (or *comedones*), acne and whiteheads. Also considered here are dandruff and eczema.

Seborrhoea is concerned with the production of sebum, usually implying that too much sebum is being produced. This gives the skin and hair an oily or greasy look and creates make-up problems. The condition can lead to *blackheads* which are caused by the accumulation of sebum and keratin fragments blocking the entrances of hair follicles. Chemical changes cause this mixture of sebum and dead skin cells to become black. Blackheads particularly affect teenagers because changes in hormone balance (i.e. relative amounts of hormones produced) cause the increase in sebum production. It is also suggested that frequent cleaning away of sebum from the skin encourages greater production or release of sebum; also, apparently some externally applied chemicals encourage the sebaceous glands to become more active.

Acne means inflammation of the sebaceous glands and commonly develops from comedones. The resulting skin lesions are usually reddened papules (raised areas of the skin or pimples) or pustules (pimples containing pus). In more troublesome cases there is appreciable swelling of the sebaceous glands giving rise to sizeable cysts beneath the skin.

Small pimples of pearl white appearance consisting of small firm lumps beneath the skin are called *milia.* They are due to sebum being trapped with no access to the surface or, some would say, they are dense keratin cysts. They frequently occur in the delicate area beneath the eyes and are also commonly known as whiteheads.

Inflammation of the skin is *dermatitis* and it may be *contact dermatitis* due to contact with irritant chemicals or other external causes, or it may result from some cause within the body.

Eczema is a non-contagious dermatitis the effects of which include irritation, vesicular lesions and scaling of the skin. It may be of internal origin or due to external causes.

Other non-infectious skin disorders are *psoriasis* in which scaling of the skin occurs, and *urticaria* in which raised, reddened and itchy areas (*wheals*) occur for a short time, often as the result of foods being eaten to which the person is allergic.

It is normal for cells of the horny layer of the skin to be continually shed, but this process is not usually noticed. Visible flaking of skin from the scalp is called *dandruff* and may be associated with bacterial infection. Otherwise, scaling of the skin can result from alkalis in soap and from strong detergents.

Bacteria

So tiny are *bacteria* that they are only visible with powerful microscopes and so they are called *microbes* or micro-organisms. They are alive and made of protoplasm but consist of only a single cell, and no clear nucleus can be seen in a bacterium.

Most bacteria are harmless; some are vital to us, breaking down waste food in the large intestine, but others are harmful (*pathogenic*).

Bacteria are described as *cocci* if they are round, *bacilli* if they are rod-shaped; others have the shape of curved rods or of a comma, or are coiled (*spirillae*). Also some cocci are always found in clusters (*staphylococci*), others form chains (*streptococci*), while others occur in pairs. The different types are shown in Fig. 22.2.

Because bacteria are living things they must use oxygen for respiration and give out carbon dioxide. They may have things like tails that can move so as to propel the bacteria and they all feed, grow and multiply. Like other simple organisms they can reproduce very rapidly.

Most bacteria must obtain dissolved carbohydrate from dead matter, plants, or animal tissues which they infect. So they secrete enzymes that digest the food material and the dissolved food passes into the bacterium cell.

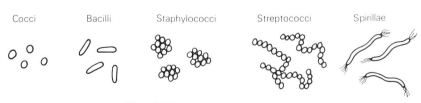

Cocci   Bacilli   Staphylococci   Streptococci   Spirillae

Fig. 22.2   Types of bacteria

The harmful effects of bacteria on the body are mostly due to poisonous chemicals (*toxins*) released into the body's tissues.

The *staphylococcus aureus* bacterium is normally found on the human skin and often in the nose of a healthy person. However, under 'suitable' conditions, it can multiply enough to be quite harmful. It is one of the commonest causes of pus-forming infections such as boils and carbuncles. It can infect wounds and can sometimes be responsible for food poisoning.

Another common pathogenic bacterium is *streptococcus pyogenes*. This is found in the throat even in a healthy person; it is commonly the cause of throat infections.

Pathogenic bacilli include the *tetanus* bacterium found in soil, the *salmonella* bacterium (normally found in the intestines of animals) that can cause food poisoning, and certain bacteria that normally live harmlessly in the human large intestine. But these 'harmless' bacteria can cause disease if, for example, they find their way into the urinary system.

Bacterium spores

Some bacteria, especially bacilli, can survive adverse conditions by forming *spores*, i.e. each bacterial cell forms a hard wall around itself which will keep it safe and moist inside until conditions for it improve.

The tetanus bacterium is an example.

Viruses

These are far smaller even than bacteria and there is some argument about whether they are living things or just inanimate chemicals. A *virus* on its own appears to be a chemical molecule with no signs of life, but once it reaches a living cell it makes the cell produce more viruses like itself, disrupting the cell's normal metabolism in the process.

Soon the attacked cell is caused to break open, liberating the new virus particles, i.e. 'breeding' of the virus has taken place.

Viruses are responsible for diseases such as the common cold, influenza, chicken pox and measles. Cold sores that appear as blisters on the lips from time to time are due to viruses which, once in the body, produce cold sores whenever the body's defence system (its resistance to disease) against such organisms is weakened. Warts are mostly caused by viruses too.

The body's defence against infection

First, bacteria may be killed by bacteria-killing enzymes found in the mouth, nose and some other parts of the body. Even sweat and tears have antibacterial properties.

Next, a bacterium or virus that might enter the body must penetrate the horny layer of the skin or, for example, the mucus coating over the inner surfaces of the mouth and nose and throat. If it succeeds and gets into body tissue, it will then find that in any body tissue there are cells (macrophage cells) produced by the

reticulo-endothelial system which will attack. Also the intruding organism will gradually become surrounded by large numbers of the various white blood corpuscles.*

A macrophage cell spreads around a foreign body like a bacterium so that the bacterium finds itself sealed inside the surrounding cell in which it will be killed and digested as shown in Fig. 22.3 (overleaf).

Most of the white blood corpuscles also have this ability to 'eat' bacteria; others are concerned with the produc-

---

*The white corpuscles are attracted to the area under attack by chemicals released by the damaged cells and by the invading bacteria themselves.

267

tion of chemicals, called *antibodies*, that will attack the bacteria or remove the particular toxins created by the bacteria. The type of antibody that is produced must suit the bacteria causing the infection. Some white cells are killed, but the invading bacteria or viruses (hopefully) will finally be killed and removed by the phagocytic cells or, in some cases, removed in pus that pushes out through the body surface.

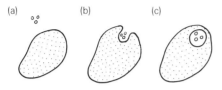

Fig. 22.3 Phagocytosis — bacteria being eaten:
    (a) a phagocytic cell, i.e. a macrophage cell;
    (b) the cell engulfs the bacteria;
    (c) the bacteria are captured and will be killed

Two further mechanisms assist in the defeat of the enemy. One is the production to some extent of chemicals that act against a wide variety of different bacteria or viruses, *interferon* being a chemical that overcomes many virus infections. The other mechanism is the lymphatic system. Lymphatic drainage of tissues is always increased when there is inflammation and there will be a simultaneous increase in escape of fluid from blood vessels. Some bacteria may enter the lymphatic system before they are killed. Then the lymph nodes, it seems, hold up the bacteria, and macrophage cells there attack the bacteria.

Thus we see that bacteria are opposed by bactericidal enzymes, difficulty of entering the body, macrophage action, antibodies, other chemicals and by lymphatic system action. Viruses are opposed in similar ways.

Spread of infection

The usual methods by which bacteria and viruses are delivered to their hosts are:

  (a) Droplet infection where minute water droplets containing the germs are carried in the air from the mouth or nose of an infected person. Colds and influenza in particular spread like this.

  (b) By contamination of food or drink with faecal material (even quite invisible amounts) from infected animals or people. Food poisoning can occur in this way.

  (c) By body contact with direct transfer of infected saliva, faecal material, pus or skin cells.

  (d) Transfer by instruments, towels or salon furnishings that come into contact with clients. Verrucas (see p. 271) are due to viruses easily picked up when bare feet touch an infected floor.

  (e) Self-infection by bacteria normally present on the body, for example, when a wound provides a very favourable breeding ground for staphylococcus bacteria.

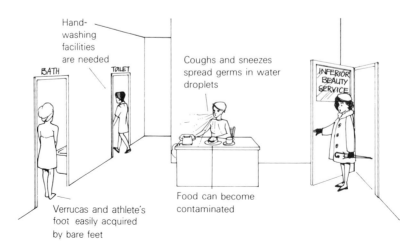

Fig. 22.4 The spread of infection in a salon

(f) By tetanus bacteria present in the soil which can therefore be present in dirt on the floor. A large number of such bacteria can enter a wound with dreadful results, usually in cases where the wound is deep.

A summary of these causes is shown in Fig. 22.4.

Fungus infections

A *fungus* is a simple plant containing no chlorophyll so that it has to obtain its energy from complex chemicals taken from living or dead materials on which it lives. (See Fig. 22.5(a).)

Athlete's foot and so-called *ringworm* are both due to infection of the skin with a fungus. Infection is usually introduced when fragments of fungus-infected skin or hair enter a scratch or cut in the skin. (See Fig. 22.5(b)).

Bare feet in communal bathrooms and changing rooms are an invitation to athlete's foot.

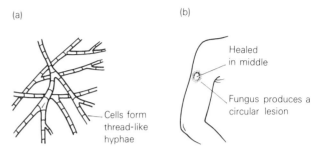

Fig. 22.5 A fungus, e.g. ringworm:
(a) the structure of a fungus;
(b) appearance of a ringworm infection

Insect pests

Parasites are living things that live on and obtain their food from other living things. Fleas and lice are blood-sucking parasites which bite through the skin of the infected person and so cause uncomfortable irritation but also provide breaks in the skin through which germs can enter.

Antiseptics and disinfectants

A *disinfectant* kills micro-organisms if it is used in sufficient concentration and allowed sufficient time to work.

An *antiseptic* at least prevents multiplication (i.e. produces sterilisation) of the micro-organisms and *may* kill them.*

Antiseptics are generally milder than disinfectants and are frequently used on the skin.

A fairly strong solution of alcohol (ethyl or isopropyl) in water can be used to disinfect tools, or a solution of QUAT (a quarternary ammonium compound), or sodium hypochlorite, or formalin (which contains formaldehyde, also called methanal). Fig. 22.6 shows a cabinet used for disinfecting small items of equipment with formalin.

For cleaning floors, walls and working surfaces various commercial products are suitable; they often contain cresols or phenols or chlorinated xylenols.

Fig. 22.6 A cabinet for chemical sterilisation

For antiseptics a weaker alcohol solution can be used to wash the skin. Hydrogen peroxide solution of suitable strength or cetrimide are other possibilities.

---

*Alternatively the term *antiseptic* is applied to a disinfectant for use on living tissue, e.g. the skin.

Above all else, we must include soap and water or mild detergent and water as a means of ridding the skin of unwanted bacteria. Washing them off is simple and can be highly effective.

Hexachlorophene is an antiseptic often included in cosmetics.

Many chemicals that kill bacteria and fungi are ineffective against viruses. Formaldehyde, however, does render viruses non-infectious.

With all antiseptics, particularly disinfectants and, in fact all chemicals, it is essential to use them only as directed by the suppliers. Phenol and formaldehyde are very toxic. With such chemicals it is often necessary to wear protective clothing and goggles and to ensure very good ventilation (a strong draught) to blow away harmful vapours.

Other methods of sterilisation

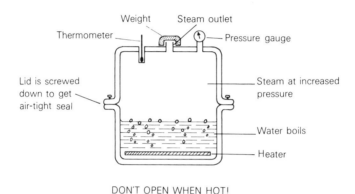

DON'T OPEN WHEN HOT!

Fig. 22.7 An autoclave

If germs are subjected to a sufficiently high temperature (70°C or more), they are killed. This is because the germ protein is coagulated, i.e. is changed from fluid to semi-solid (like boiling an egg).

A higher temperature is needed to kill spores if these are produced by the germs. Dry heating in an oven can kill germs but the temperature required (for example, 150°C for 1 hour) is too high for items such as those made of plastics, and damp items may be slow to reach the necessary temperature.

If the bacteria are heated in water or steam they are more easily killed. Boiling in water for 5 minutes kills bacteria and is suitable for cotton towels and sheets as well as for metal items (although sharp tools like scissors may be gradually blunted).

For moist heat at a higher temperature an *autoclave* is used, although a domestic pressure cooker is an alternative if only a small size is required. The principle used here is that the boiling point of water is raised if the pressure on the water is increased above the normal

atmospheric pressure. This pressure increase can be achieved simply by not letting the steam escape from the container in which the water is boiled. The pressure builds up and is controlled by having a weight of chosen size that rests over the opening from the autoclave to the outside air (see Fig. 22.7). A temperature of 120°C is commonly employed (requiring a pressure of 1 atmosphere greater, 76 cm of mercury, than the normal outside air pressure). Bacteria and spores are killed.

Note that the steam should first be allowed to drive out all the air from the autoclave. Otherwise the expected temperature will not be reached.

Ultra-violet radiation of short wavelength (UVC) is germicidal, and low-pressure mercury vapour tubes, made of a glass that allows UVC to pass out from the tube, emit UVC (in fact 254 nm) and comparatively little radiation of longer wavelengths. Fig. 22.8 shows a cabinet fitted with such a tube which can sterilise equipment.

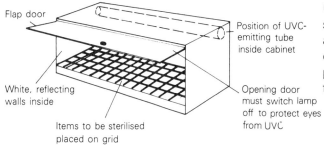

Flap door

White, reflecting walls inside

Items to be sterilised placed on grid

Position of UVC-emitting tube inside cabinet

Opening door must switch lamp off to protect eyes from UVC

Fig. 22.8 An ultra-violet sterilising cabinet

Unfortunately, because radiation travels in straight lines, some parts of the items being sterilised will lie in shadow areas in spite of reflection of the UV from the walls of the cabinet. For this reason, and because UVC cannot penetrate dirt films, the items concerned should be thoroughly washed first.

More skin problems

Perhaps the least of these is the development of *freckles* which are small areas coloured brown by melanin. This effect occurs in the horny layer of the epidermis.

Vitiligo is an affliction where patches of skin are quite devoid of pigment. *Chloasma* is rather the opposite because it involves areas, most often on the face, of brownish pigmentation.

Areas of redness, particularly on the cheek, are caused by broken blood capillaries.

These, as well as other non-contagious skin conditions such as some kinds of dermatitis and moles, are of concern to the beauty therapist. Information regarding these is readily available from medical literature as well as from well-known beauty therapy books.

Warts

Also called *verrucas* (or *veruccae, warts* are usually caused by viruses and are more or less contagious.

The virus makes the prickle-cell layer of the skin undergo mitosis sufficient to cause a small solid outgrowth of the skin called a wart.

There are several types of wart (see Fig. 22.9). The threadlike (*filiform*) and soft warts are most often found on the neck where rubbing of the clothes or jewellery occurs, and the common wart is found mostly on the backs of the hands. These warts are contagious to a small extent and can spread on the same person, but this does not happen much. If they are not removed, the warts tend to disappear spontaneously after a year or so.

In contrast, warts under the feet (*plantar* warts or verrucas) can be very uncomfortable. These verrucas consist of a mass of dense, moist tissue which is kept pushed into the foot by the body's weight pressing the foot down. They are easily acquired by walking bare-footed on the floors of swimming pools, bathrooms, etc.

The *seborrhoeic* or *senile* wart is not due to a virus and is not at all contagious.

Warts can be removed but medical advice should be obtained first.

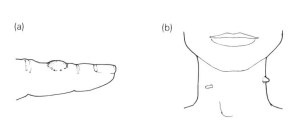
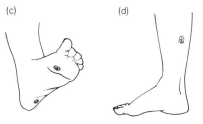

(a) (b) (c) (d)

Fig. 22.9 Some types of wart: (a) a common wart; (b) filiform warts; (c) plantar verrucae; (d) a seborrhoeic wart

Washing

Washing with soap and water or mild detergent and water will loosen and wash away dirt, which contains dead skin, sweat and sebum together with cosmetics, dust and fluff from clothing. Without the soap or detergent the oils in which the other waste is mixed would not be easily emulsified and removed.

Not only does washing make us look clean, but the removal of dirt will minimise the body odour that soon develops as bacteria act on the dirt.

Washing and bathing can also be enjoyable, and warmth from the water can be soothing and relaxing.

Soaps and detergents

The success of *soaps* and *detergents* for washing purposes and for use in cosmetics is due to their being emulsifying agents.

Soap

As pointed out in Chapter 18, a soap is formed when a strong alkali reacts with a fat. For example:

Sodium hydroxide + Glyceryltristearate
   = Sodium stearate + Glycerol*

A soap-forming process is called *saponification*.

Note that if some alkali remains in the final product it can be the cause of skin irritation. This is more likely to happen with cheaper soaps.

The structure of the soap molecule

This is illustrated in Fig. 22.10 by the sodium stearate molecule. †

The stearate group $C_{17}H_{35}$

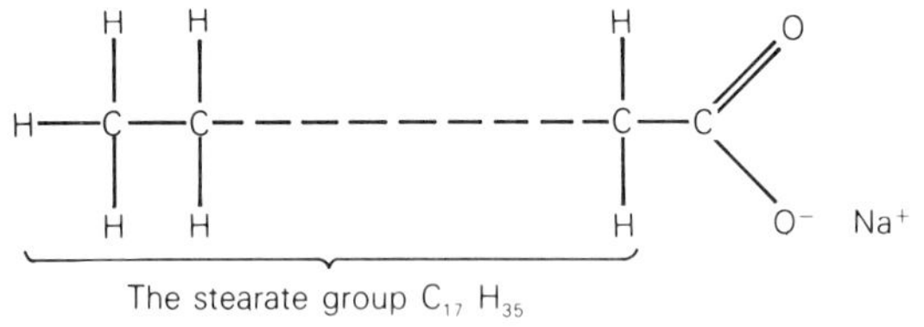

Fig. 22.10   The structure of the sodium stearate molecule

---

*Examples of the fats (and oils) commonly used are tallow (beef and mutton fats), coconut oil, and palm oil.

The main glycerides that these contain are, respectively stearic and oleic, lauric, palmitic and oleic.

During the manufacturing process common salt is added to the soap—water—glycerol solution and this precipitates the soap.

Sodium stearate, oleate and palmitate are suitable for toilet soaps and soap powders. A soft soap which might be used in a shampoo can be made, using a different manufacturing process, by reacting a vegetable oil such as palm oil with potassium hydroxide.

† The oxygen atom (O) and the sodium atom (Na) at the right-hand end of the molecule in the diagram are not linked by sharing an electron as might have been expected but the oxygen atom has taken the weakly held electron from the sodium atom so that we now have an O^- ion and an Na^+ ion. The sodium ion now holds on to the rest of the molecule by electrical attraction of the opposite charges. It is the presence of these charges that makes one end of the soap molecule attractive to water molecules. The other end of the molecule will mix readily with oil molecules.

The cleaning action of a soap

Soap 'dissolves' in water by forming an emulsion as explained in Chapter 9 (Fig. 9.18). Grease from the skin is able to dissolve by its molecules finding places for themselves among the stearate ends of the soap molecules as shown in Fig. 22.11. Dirt held in the grease is therefore freed.

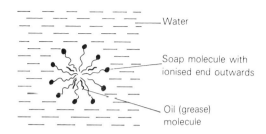

Fig. 22.11   Dissolving of grease in soapy water

Synthetic detergents

A *detergent* is a surface-active agent that is used for cleaning and so soap is really a detergent. However, the term is usually reserved for *synthetic* detergents, not derived from animal or plant products but often from petroleum products.

A typical detergent has, like a soap, a molecule with one end ionised so that it is attractive to water while the other end is not.

Detergents can differ in several ways from soaps but, in particular, detergents do not form scum when used in hard water.

A good example of a detergent is shown in Fig. 22.12.*

Triethanolamine lauryl sulphate (or TLS) has the advantage of being a colourless liquid, soluble in water, and suitable for clear liquid shampoos as well as being very popular for bubble baths.

Surface-active agents can also be useful for obtaining foams by reducing the surface tension of water.

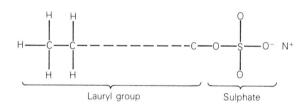

Fig. 22.12   A detergent molecule — sodium lauryl sulphate

Shampoos

A *shampoo* should remove grease and dirt from the hair and scalp, but only surface cleaning is necessary. Normally the shampoo will remove some of the natural oil from the hair and so will affect its texture. Ideally it will leave the hair not only clean but with an attractive lustre, fragrance and texture.

Shampoos may be clear liquids, lotions, pastes, gels, aerosols or dry products.

An example of a clear liquid shampoo is one consisting of TLS (the detergent) dissolved in water, plus a foaming agent (to improve the lather) and perfume.

Various additions can be made, particularly in the case of lotions, such as mineral oil, lanolin or glycerol to remain on the hair as an emollient. Egg, beer, lemon and proteins are other conditioners (to improve hair texture, lustre or manageability). Colouring, antiseptics and preservatives may be also added.

*It is a cationic molecule, i.e. the electrically attracted ion is positive charged. It is used as a detergent in shampoos and bubble baths and as an emulsifier in many cosmetics. It is not very soluble in water.

Dry shampoos

A *dry* absorbent powder mixed with a mild alkali is used when a person must avoid washing the hair because of some ailment. When rubbed into the hair, the powder collects the dirt and grease from the hair and scalp. After sufficient time for this to act, the powder is combed or brushed out.

Also, liquid shampoos not employing water can be called 'dry'. An organic solvent is applied to the hair and this dissolves the grease. The liquid containing the dirt can then be rubbed off with a towel.

Antidandruff shampoos

These are most often opaque lotions containing either selenium sulphide or zinc pyrithione. These ingredients remove the loose keratin scales from the scalp so that they are washed away. Regular use is needed in order to avoid any signs of the dandruff.

Bath salts and bubble bath

When we want foam and fragrance in the bath we like a *bubble bath*. The important ingredient of most bubble baths is a foaming agent (that reduces the surface tension of the water); a separate detergent (for cleaning the body) may be added if the foamer is not a satisfactory detergent in itself. The detergent has an advantage over soap in that it does not produce a scum and so there is no tide-mark left on the bath.

A liquid bubble bath may consist simply of TLS, which is a detergent that foams well, plus a chemical to ensure that a good foam is produced even if soap is added. Add to this a little perfume and colouring; the remainder of the bubble bath is water. (See Fig. 22.13.)

Bath salts are for water-softening or for adding colour and fragrance to the bath water. In Chapter 18 it was explained that sodium carbonate (washing soda) will precipitate the hardness in water. In fact, crystals of sodium carbonate can be used as bath salts, but this practice makes the water alkaline so only a little should be used. Some other sodium salts may be used instead. Sodium chloride crystals with some perfume and colour added can be used, but these crystals do not soften the water.

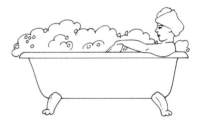

Fig. 22.13 Foaming is usually considered more important than good detergent action

Deodorants and antiperspirants

The main cause of unpleasant body odour is normally the bacterial decomposition of sweat, particularly apocrine sweat. This occurs particularly in the armpits.

A *deodorant* is a skin application designed to stop smells that could be produced by sweat, achieving this by simply preventing the bacterial decay. In other words, the deodorant is a germicidal chemical or antiseptic such as hexachlorophene.

An *antiperspirant* is designed to stop the release of sweat on to the skin surface. Aluminium chloride can be used (it is soluble in water) and it simply blocks the pores. Aluminium chlorhydrate is less likely to harm clothing.

It is common to combine a deodorant and an antiperspirant.

Such cosmetics must be easy to apply and non-sticky, and should not stain clothing or irritate the skin. The desired effects should last for several hours.

The active ingredient can be dissolved in an oil in water emulsion or can be presented in aqueous solution that can be sprayed on.

Choice of clothing

The heat-insulating properties of various clothing materials have been considered (in Chapter 10, p. 106). What is also important is the absorbency of such materials and the ease with which they can be cleaned.

Loosely constructed fabrics are the more absorbing. Wool fibres absorb moisture quickly and can hold a lot of moisture, linen absorbs water quickly and dries quickly (good for towels), and cotton absorbs quickly but does not dry as quickly as linen.

Smooth surface fibres, like those of linen, silk and rayon, tend to keep clean and can be washed easily. Cotton soils easily due to the rough surface of its fibres, but it can be washed easily and can be boiled.

Wool fibres collect dust and bacteria and soon become dirty so that frequent cleaning is necessary.

Washing and toilet fittings

The familiar wash hand-basin is illustrated in Fig. 22.14 (a) and (b).

The glazed stoneware basin is strong and easily cleaned with a non-abrasive cleanser such as a detergent. It is a poor conductor of heat and looks attractive. It is easily chipped so that stainless steel is sometimes chosen instead.

The basin is fitted with a 'trap' beneath it to stop unpleasant smells and airborne germs from entering the

clinic from the waste pipes. The trap is described as an S-bend or P-bend according to its shape and is fitted with an 'eye' that can be removed to clean out hairs and other debris that accumulates in the trap. Alternatively, a bottle-trap is fitted and the lower part of this can be screwed off.

The same trap principle is used in the WC pan.

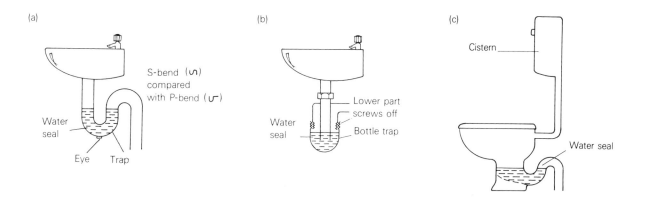

Fig. 22.14   Washing and WC fittings: (a) a hand-basin with an S-bend;   (b) a hand-basin with bottle trap;   (c) a WC pan and cistern

SUMMARY

- Inflammation is the redness, swelling and often pain caused by injury or ailment.

- Abscesses are lesions where pus is produced.

- Clotting of blood involves the conversion of fibrinogen to threads of fibrin and is caused by chemical released by blood platelets damaged by the bleeding.

- Seborrhoea means excessive sebum production.

- Blackheads (comedones) are caused by blocking of hair follicles.

- Whiteheads are buried comedones or milia.

- Eczema is a dermatitis characterised by skin irritation, vesicular lesions and scaling.

- Red blotches on the skin may be urticaria.

- Contact dermatitis is caused by irritants such as chemicals.

- The important types of bacteria are cocci, bacilli, spirillae, staphylococci and streptococci.

- Viruses alone behave merely as chemicals but in contact with living cells they make cells produce more viruses, and the cell suffers.

- Food poisoning and most skin infections are caused by bacteria.

- Colds, influenza and measles are caused by viruses.

- Germs are attacked by phagocytic (macrophage) cells present in the invaded tissue and also by the arrival of white blood corpuscles.

- Infection can be spread by droplets, contamination of food, body contact, dirty instruments, dirt on floor, etc.

- Examples of fungus infections are athlete's foot and verrucas on the foot.

- Warts, including verrucas, are mostly caused by viruses.

- Plantar warts and athlete's foot are easily acquired from floors of bathrooms, etc.

- Antiseptics stop bacteria multiplying.

- Disinfectants kill bacteria, i.e. they are bactericides.

- Viruses are killed by some chemicals.

- Formaldehyde vapour is used as a disinfectant.

- Hexachlorophene is an antiseptic used in some cosmetics.

- Germs can be killed by dry heat, by boiling for 5 minutes or by use of an autoclave.

- UV radiation, especially UVC, has germicidal properties.

- Soap 'dissolves' in water to form an emulsion in which grease from the skin can dissolve so that dirt is freed.

- Detergents behave like soaps but produce no scum.

- Hair shampoos often consist of solutions of detergents.

- Antidandruff shampoos may contain selenium sulphide or zinc pyrithione.

- Bath salts can soften the bath water, or add colour, or fragrance.

- Deodorants are usually antiseptics that prevent bacterial decomposition of sweat.

- Antiperspirants block the sweat pores. Aluminium chlorhydrate can be used as the active ingredient.

- Sinks and WCs could be a source of smells and germs but for the use of traps that provide a water seal.

EXERCISE 22

1. What is meant by *inflammation*?

2. What is the function of *fibrinogen*?

3. Distinguish simply between *staphylococci* and *streptococci*. (CGLI)

4. Name two diseases caused by viruses.

5. In what ways can bacterial infection be spread from person to person?

6. Distinguish between *antiseptics* and *disinfectants*.

7. Name two antiseptics.

8. Define the word *detergent*. (CGLI)

9. Name an active ingredient used in antidandruff shampoos.

10. Name an active compound in: (a) an antiperspirant, (b) a deodorant. (CGLI)

11. How do deodorants work?

12. What is the purpose of an autoclave? (CGLI)

13. Where is an S-bend found in the salon, and what is its function? (CGLI)

14. State *two* disadvantages associated with ultraviolet sterilising cabinets.

15. With what are salmonellae frequently associated? (CGLI)

16. Name a disease of the skin caused by a fungal parasite. (CGLI)

23

FIRST AID

WHAT IS FIRST AID?

First aid only

First aid means the first help given to an injured person. It is accepted that anyone confronted with an injured person should offer sufficient help to save life and prevent further injury. These are the aims of first aid. Treatments that go beyond these aims will mostly be the concern of doctors or other appropriately trained medical people. An unqualified person undertaking more than first aid might even come into conflict with the law.

First aid regulations

In the UK there are now regulations under the Health and Safety at Work Act which require even small establishments to have one or more first aid boxes, to provide information for employees, to keep records and to appoint a person to be in charge of first aid matters. The exact contents required in first aid boxes are specified. There are further requirements for establishments that employ many people.

HANDLING THE EMERGENCY

Approaching the injured person

It is important that you, the first aider, show no signs of panic, but you should approach the injured person *without delay*. Speak *confidently, clearly* and with no sign of alarm. Say to the injured person: 'Lie still. Where are you hurt?' Your manner must be *reassuring*.

Hazards that might cause further injury, perhaps broken glass or furniture about to fall, should be removed immediately.

Normally you *do not move* the patient until you have made certain that doing so will not lead to further injury.

Finding out what is wrong

When the patient is conscious ask where she is hurt or what has happened. This leads to a *quick investigation* of the injuries and a check over the body for any bleeding or broken bones. When the patient is unconscious the quick investigation involves checking that the person is breathing normally, and next looking for any severe bleeding and then for any broken bones. Deciding what has happened may be helped by information from bystanders.

Life-saving treatment must begin at once, but now is also the time to send for *medical assistance* if necessary.

After these priorities have been dealt with a *second, more thorough check* for injuries can be made and the appropriate treatments given.

Sending for help

For all but very minor emergencies an ambulance will be sent for, although in exceptional circumstances the patient might be taken by car to a doctor or to a hospital. To send for this help you must choose an intelligent-looking person (who is also required to remain calm and collected) to convey your message, usually by telephone.

The message must be brief and clear, and it should state what help is needed (usually an ambulance, perhaps a doctor if possible, sometimes the fire service as well), the

exact location of the accident and a brief but clear indication of the apparent or suspected injuries, for example, 'broken leg' or 'severe bleeding from the arm'. The messenger should report back to you.

It may also be necessary for someone to wait where the ambulance can be quickly directed to find you.

A summary is given in Fig. 23.1.

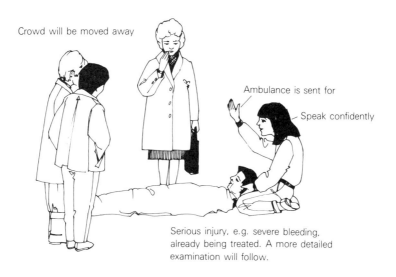

Crowd will be moved away

Ambulance is sent for

Speak confidently

Serious injury, e.g. severe bleeding, already being treated. A more detailed examination will follow.

Fig. 23.1   Taking control of the situation

Serious injuries

Death may quickly result from several causes which include:

Cessation of breathing
Severe loss of blood
Damage to internal organs
Poisoning
Extensive wounds
Surgical shock

Burns or broken bones and electric shock can, for example, cause some of these effects; or a mishap inside the body may be the cause, such as a stroke or a heart attack.

Injuries causing any of these conditions are serious. Inadequate breathing and severe bleeding should be one's first concern as explained below.

Lesser injuries

In this category we can include a nose-bleed, fainting or cramp in a muscle, as well as small cuts, bruises, grazes and small burns. These usually require no professional help.

TREATMENTS

Quality of treatment

In addition to receiving treatment for injuries the patient should be talked to so that she is reassured and comforted. Her injuries should be discussed sufficiently for this purpose. Often too the patient is worried as much about getting a message to relatives or friends or about the well-being of personal belongings and these problems should be remedied. Also make the patient as *comfortable* as possible.

It is possible that an incompetent handling of an emergency in a salon could reflect badly on the prestige of the establishment. Competence is obtained by experience, by study and by advance planning, supplemented by common sense, alertness and other personal attributes. Attending a course in first aid is a good idea.

The brief details given below should be regarded as the minimum knowledge required by the beauty therapist who is in charge of a clinic (even if only temporarily). The signs and symptoms and treatments really need to be memorised.*

Breathing absent or difficult

If breathing stops, for example due to an electric shock, then the blood, and so also the brain, can become short of oxygen. As a result the brain will cease to function. Equally, if the brain or heart stops working properly due to accident or illness, this may affect the working of the lungs.

Absence of breathing requires *artificial ventilation* to be started immediately. A solid obstruction like a sweet lodged in the mouth, throat or windpipe can make breathing very difficult, so that the person chokes, splutters, coughs and shows great distress. The breathing might be noisy or sound bubbly. The patient may feel better if she takes long, deep breaths. Coughing may release the obstruction. If the situation becomes desperate and the object can be seen, an attempt should be made to remove it, possibly by hooking your finger around it. For something deeper down, sharp blows can be applied between the shoulder blades with the patient lying on her side. The same position would be chosen if the problem were choking by liquid in the throat. If the treatment fails and the patient becomes limp or unconscious, then artificial ventilation should be started even if there is still an obstruction.

Other causes of breathing difficulty include suffocation by something over the mouth and nose or pressure on the chest or sudden illness like a heart attack. Smoke or fumes entering the lungs, of course, can prevent successful breathing.

---

*Signs are visible indications of illness or injury. Symptoms are invisible indications which are felt by the person concerned.

Mouth-to-mouth resuscitation

This is the recommended method of artificial ventilation. For success it is essential that it is started without delay, pausing only to ensure that the mouth and throat are not obstructed, and with the patient lying on her back as shown in Fig. 23.2. Close the patient's nostrils, by pinching with your thumb and fingers, tilt the head well back, take a deep breath, seal your mouth over her open mouth, and blow strongly and steadily until you see her chest rise. Lift your mouth away from her to breathe out. Watch her chest fall, take a further deep breath and repeat. Three or four inflations should be given as quickly as possible. If the heart is beating normally continue your mouth to mouth ventilation at the rate of 16–18 times per minute. If the patient vomits then quickly turn the head to one side so as to drain the fluid from the mouth.

Closing the patient's mouth and blowing through her nose is an alternative method.

Mouth-to-mouth/nose ventilation should be continued until natural breathing begins or until medical help arrives.

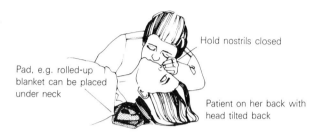

Hold nostrils closed

Pad, e.g. rolled-up blanket can be placed under neck

Patient on her back with head tilted back

Fig. 23.2 Giving artificial respiration to resuscitate a patient

Heart stopped

If your mouth-to-mouth ventilation on its own is unsuccessful check to see if the patient's heart has stopped beating. This is best done with three fingers to check for the carotid pulse, as shown in Fig. 23.3. If you are quite sure there is no pulse then start external chest compression. This is a skill that should be learned but briefly it involves pressing with controlled pressure upon the breastbone as shown in Fig. 23.4.

Kneel to the side of your patient and with your arms straight press down on the lower half of the breastbone. Relax to release the pressure. Fifteen compressions should be given at the rate of eighty per minute. If alone you alternate between mouth-to-mouth ventilation and external chest compression at the ratio of two ventilations to fifteen compressions.

As soon as the patient has a heart beat and is breathing place her in the recovery position (see Fig. 23.8).

Note: While mouth-to-mouth ventilation can be given to assist a patient's breathing, external chest compression must be stopped at the first sign of a heart beat.

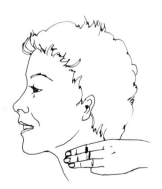

Fig. 23.3 Taking the pulse at the neck

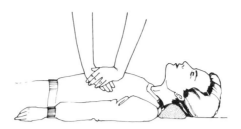

Fig. 23.4 External chest compression

Severe haemorrhage

Haemorrhage means 'bleeding' and when severe, from a large vein or artery, it must be stopped urgently. The body contains about six litres of blood, and loss of much more than half a litre can soon cause loss of consciousness followed by death.

Blood from a vein flows out steadily and may be dark in colour. From an artery it may spurt out and it will be the familiar bright red.

The action to be taken is first to lay the patient down (to reduce the heart beat), then the limbs can be raised to concentrate blood in the heart and body. Holding together the edges of a gaping wound can reduce bleeding, but the normal procedure is to place a large dressing or small towel as a pad over the wound and hold it or bandage it firmly so as to apply pressure on the wound area (Fig. 23.5). This pressure must not be relaxed for ten minutes. Care should be taken that the dressing does not apply pressure all around a limb because this might completely stop the blood supply to the limb beyond the dressing.

In addition to bleeding from external wounds blood may come from the mouth due to injury or disease affecting the stomach or lungs. Blood from the lungs might be bright and frothy while that from the stomach would be very dark in colour. In some cases of internal bleeding, no blood is seen. The condition is very serious. Many of the symptoms are those of severe shock (see below) and hospital treatment is urgent. The patient should be laid down with head to one side and kept low. The only possible first aid treatment is treatment for shock.

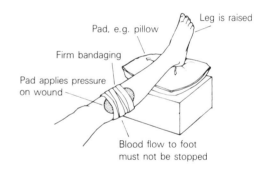

Fig. 23.5 Stopping severe bleeding

Shock

Fright or fear can make a person suddenly feel faint. She feels light in the head, sick in the abdomen, shaky in the limbs. She looks pale, and the skin is cold and perhaps wet. However, this emotional or nerve shock is not medically serious and recovery can be quick.

True shock or injury shock comes on soon after a serious injury and can be serious and fatal. Haemorrhage and burns particularly give rise to such shock. Broken bones frequently cause haemorrhage resulting in severe shock.

Shock is associated with loss of fluid from places where it should be and the body reacts to this by reducing the blood supply to the skin. The skin consequently is pale or grey, cold to the touch and wet and clammy due to a cold sweating. Breathing is likely to be abnormal and the pulse weak and fast. The patient may be restless and complain of thirst. She may be unable to think clearly as if she were drunk. The patient may lose consciousness.

When shock is present or is expected, not only should the injuries responsible for it be treated but the patient should be treated for shock. She should be made to lie down preferably with feet raised slightly, the head low perhaps on a pillow for comfort and her (his) head turned to one side (in case of vomiting). The patient should be kept warm with blankets or coats. Tight clothing should be loosened. Keep calm and reassure the patient. Do not give her any drinks. Severe shock requires hospital treatment.

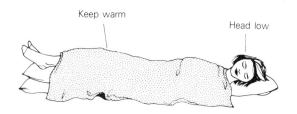

Fig. 23.6 Treatment for shock

Treating wounds

With any wound, priority must be given to stopping the bleeding if this is serious and the procedure in this case has already been discussed. Otherwise the wound is dressed as follows.

The skin surrounding the wound is washed using swabs of cotton-wool or a clean cloth wetted with tap water or with a solution of mild antiseptic such as cetrimide which does not sting. Soap and water is an alternative. This washing removes blood and dirt near the wound, and its main purpose is to reduce the risk of infection of the wound. A small wound on the hand or arm could of

course be washed under a running tap. A dry swab or cloth is used to dry the area. The wound can now be covered with a dry, clean, preferably sterile dressing and is bandaged to hold the dressing in place. In the case of a small cut, the patient will not need expert medical attention, and so the dressing you apply is not merely a temporary measure and your professional prestige requires that the standard of your dressing is equal to that of your beauty therapy. Do the job neatly. Fig. 23.7 illustrates the dressing and bandaging of a finger.

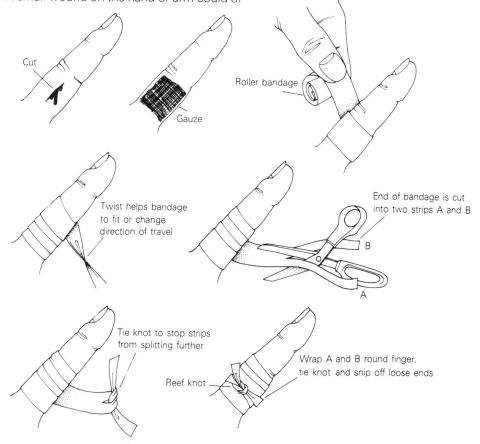

Fig. 23.7 Applying a dressing to a finger wound

A wound that is deep or whose edges can gape (so that stitching could be required) needs proper medical attention. A sterile dressing should be applied and the patient taken to hospital or a doctor. Even a small wound should be examined by an expert if skin has been lost to a depth of a millimetre or more.

Because escaping blood washes wounds, it is not usual to wash a wound, only the surrounding skin. Washing the wound could unnecessarily disturb the clotting blood. In the case of a superficial graze the wound itself can be washed. The graze is then best left exposed to the air, but a light dressing over it will probably please your patient. It can be removed when she gets home.

For small wounds a *stick-on dressing* may be used, but remember that air should be able to penetrate the dressing and water vapour escape through it so that the wound can keep dry and heal quickly.

Burns

When the skin is overheated, erythema (reddening) occurs due to increased blood supply to the area. The blood vessels there are dilated and they become permeable so that fluid escapes from the blood. This can cause swelling and blistering. If burning removes some skin so that fluid escapes more easily, the large quantity of fluid lost produces severe shock when the burnt area is of substantial size.

Treatment of any burn aims at removing heat and contracting the blood vessels and this is achieved by gently applying plenty of cold water to the part. Clothing is removed as necessary. The cooling is continued for ten minutes or more, or until the pain stops. The burn is covered with a dressing as for other wounds. Any burn more than a centimetre across or that penetrates the skin should receive expert medical attention. A small, surface burn can be covered with tulle gras (a gauze smeared with medical petroleum jelly) or a proprietary antiseptic cream and dressed in the same way as a graze. If blistering is present use only a dry dressing.

With all severe burns shock must be treated while medical help is awaited. *Scalds* by hot water or steam or other hot liquids will often require rapid removal of clothing that contains the hot liquid. With *chemical burns* water is applied as for other burns and it serves to remove the offending chemical. *Electrical burns* are special because the burnt area beneath the skin is usually much more extensive than the small burn at the skin surface suggests. Expert medical advice is needed.

Splinters

Small splinters of wood, metal, etc. embedded in the skin surface can be removed with forceps or a needle point, but otherwise proper medical attention is required.

Nose-bleeding

This may occur even with no apparent cause. For treatment the patient is required to sit quietly and hold the nostrils closed for up to 10 minutes while breathing through the mouth. Swallowing should be avoided. The head is best held so as to tilt a little to prevent blood running into the throat.

Blisters

A blister caused by excessive rubbing of the skin is best just covered with a dry dressing.

Insect bites and stings

Some insects leave a small visible 'sting' which is needle-like with a sac at its end containing the stinging chemical. If so, this should be removed. Wash the area in cold water and, as with all stings, apply a cream recommended for stings (for example, an antihistamine cream).

Foreign body in the eye

A small foreign body such as an eye-lash, which finds its way on to the surface of the eye, can often be washed out. Lukewarm water is placed in an eyebath and this is held against the eye. The patient looks up into the water and blinks. Alternatively the patient could simply close the eye for a minute and hope that tears will wash out the object when the eye is finally opened.

If the foreign body can be seen and is not lying over the pupil or iris, removal can be attempted by gently lifting it out with a moistened length of twisted cotton wool. The lower eyelid can easily be moved down to show if the offending body is behind the lid, but inspection behind the upper lid is best left to the expert. Sometimes pulling the upper lid down over the lower lid enables the lower lid to wipe the body from beneath the upper lid as the eye is subsequently opened. If the body is embedded in the eye surface or is otherwise difficult to remove, then the patient must go to a doctor or hospital.

Broken bones

A client while undressing might topple over, fall awkwardly and sustain a broken arm. Outside the salon a client could slip over or be involved in a traffic accident.

A break in a bone is often called a *fracture*. As well as causing pain and shock, a broken bone can damage adjacent internal organs, nerves or blood vessels. For this reason movement of the patient may have to be avoided.

The possible signs and symptoms of a fracture are pain, swelling, loss of power (for example, in a limb), irregularity that may be felt), deformity (where an unnatural shape or position is observed) and even the sound of the broken bone ends grating together. A fracture may be confused with a sprain or strain but, if in doubt, treat it as a fracture.

The treatment for a fracture is often to await the arrival of an ambulance while giving treatment for shock and for any bleeding or wound present. If for some reason the patient has to be moved before medical help arrives, the part concerned should be supported so that the bones are held in place. A broken leg can be bandaged to the other leg (if this is not broken) with padding suitably placed between the legs to keep the injured leg in place. A broken arm can be similarly strapped to the body.

Much more could be said about fractures, but most important is the fact that a fractured spine can damage the spinal cord with the most serious consequences. If a patient seems to have hurt her back when she fell, do not move her at all. Be sure first that the spine is not fractured. Inability to move the feet or hands would be an indication of spinal injury.

Dislocations, sprains and strains

A *dislocation* occurs at a joint. The joint has come apart. One or more bones will have become incorrectly placed. A dislocation is treated like a fracture, and no attempt is made to return any bone to its normal position or indeed to move it at all. A *sprain* is a stretching or tearing of ligaments that hold a joint together and is most commonly found at the ankle joint, when the ankle becomes 'twisted'. A *strain* is damage to a muscle or tendon and is usually caused by a sudden movement. When the damage is slight the patient will soon feel able to move, perhaps with help. Sprains and strains will normally require the patient to have transport to a doctor or hospital. If in doubt, treat them as fractures.

285

Muscle cramp

Cramp of a muscle (for example, the calf muscle) is often brought on by cold, by excessive exercise, or by unusual movement. Loss of salt by sweating can encourage it. The muscle feels knotted and painful, and the condition is treated by holding the limb so as to extend the muscle as much as possible. Warmth and massage also help.

Head injuries

A blow to the head can shake the tissues of the brain and the person becomes unconscious. The person is 'knocked out', perhaps only for a brief moment but sometimes for a long time. The person may soon wake up but feel dazed and may have no memory of the accident. This injury is known as *concussion*.

A more serious condition that can follow is *compression* (of the brain). It arises when bleeding occurs within the skull and the escaped blood causes pressure on the brain. Typically the patient will become confused and gradually unconscious as the pressure increases. The pupils of one or both eyes may become very small or very dilated and finally may show no response to changes of illumination. The pulse may be slow.

The action required for a case of concussion or compression is to send for an ambulance, then keep the patient rested under quiet conditions, preferably in the recovery position (Fig. 23.8) and keep a careful record of the person's condition if the signs and symptoms change.

Of course the head injury may involve a fracture and the bones must not be disturbed. Compression could be the result of a broken bone pressing upon the brain. If a wound is present on the fractured head, there is a risk of infection reaching the brain. If some blood escapes from the nose or ear, it should be allowed to run out to relieve compression and only a light dressing is applied.

Fig. 23.8   The recovery position

Sudden illness

When a client develops a headache or begins to feel hot or cold, then she may well be aware of the cause. Anyway, unless she is in great distress, the obvious action to take is to have her driven home where a relative, friend or doctor will take care of her. The same action is appropriate if a client arrives after having a little too much alcoholic drink.

Fainting is not itself a serious condition, but certainly it can be very sudden. The brain momentarily receives too little blood and the person becomes unconscious and falls down. The condition resembles severe shock, but recovery begins immediately with consciousness soon returning. It can be caused by a frightening thought or sight. Suddenly standing up or standing up for too long can also cause fainting, especially if the person has not eaten for some time or is already unwell. Treatment requires that tight clothing around the neck be loosened and the person is allowed to lie down comfortably on her side (Fig. 23.9). If consciousness has not returned within two minutes expert medical assistance is needed. A patient feeling giddy, as if about to faint, should be encouraged to sit with her head bowed low. Lying down on a couch might be the simplest answer. Beware of the patient falling and hurting herself.

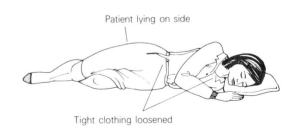

Fig. 23.9   Recovery from fainting

An *asthma attack* causes a person to struggle for breath. Such attacks are encouraged by worry and the patient needs calm, reassuring attention. Loosen tight clothing, give plenty of fresh air and make her comfortable. A sitting position may be preferred.

Other cases of sudden illness include *heart attack, stroke, epileptic fit* and *diabetic coma.* These all require an ambulance to be sent for. Heart attacks may cause severe pain in the middle of the chest perhaps affecting the shoulders, neck and arms too. The breathing may be fast and sound bubbly and the patient may look pale or even blue. The pulse is likely to be fast and irregular. Sweating may be seen.

Treatment happens to be the same as for asthma, but the heart could stop and the person will then collapse. Heart massage and artificial respiration will then be essential.

Some heart attacks are due to obstruction of the blood supply to the heart muscles (see also page 209). A similar obstruction in the brain or the breaking of a blood vessel in the brain is the cause of a *stroke.* This condition may only cause movements to become clumsy, speech to become abnormal and the person to feel giddy. In other cases the person loses consciousness. Often one side of the body is paralysed and the other side of the face remains normal. The skin may be blue, the breathing noisy and dribbling noticeable. The patient should lie down on a bed or couch with head low. Tight clothes should be loosened. Make sure that breathing is not impeded by fluid in the mouth.

A person who suffers an *epileptic fit* usually falls down suddenly, doesn't move or breathe for half a minute or more, and then produces jerky movements (perhaps biting the tongue at this stage). She then awakes and breathing is soon back to normal. Treatment should aim at stopping the patient from hurting herself as she throws her arms and legs about. Even in mild cases the person's doctor should be consulted. A serious case merits an ambulance being called.

Some people are unable to control the concentration of glucose sugar in the blood properly. They are *diabetic* and need insulin to be supplied regularly by injection or tablets to enable the sugar to leave the blood. An overdose of the drug or too much exertion or too little food eaten can lead to diabetic *coma.* The person becomes unconscious. This coma may be preceded by various symptoms almost as if she were drunk. If you think a person apparently drunk might be a diabetic, you need to look for a card which a diabetic should carry. A diabetic should also have sugar in a pocket. The first aid treatment is to give sugar if the patient is able to eat it. Medical help should be sent for.

Sudden illness of a kind not mentioned above may also involve great distress or collapse so that urgent action is needed. Regardless of the cause, medical help is sent for, the patient is made comfortable, kept warm and is not left alone. If unconscious the person is put into the recovery position. This procedure would also apply to someone who is so drunk as to be helpless or collapsed.

Poisoning

The first aid treatment for poisoning or drug overdose is to get the patient to hospital as soon as possible. Give the patient nothing by the mouth and make her as comfortable as possible.

If a corrosive liquid has been drunk, the patient can drink water to dilute the chemical but should not be made to vomit. Of course an unconscious patient is never given drink and in this case the patient is simply put in the recovery position until medical help arrives.

Emergency delivery of a baby

As with many other emergencies no problem should arise if an ambulance is sent for. Also there should be time for medical advice to be obtained by telephone or other means.

Electric shock

Before all else the electric current must be switched off. If this is not possible you must move the casualty from the electric circuit without yourself getting electrocuted via the patient. This means you must handle the patient with insulating material. After this the patient is treated for cessation of breathing and heart action, burns, shock, etc. as appropriate.

ACTION IN THE EVENT OF FIRE

A small fire can often be extinguished by quickly smothering it with a wet towel or a proper fireproof blanket or a fire extinguisher may be used, for example, a carbon dioxide type extinguisher as illustrated in Fig. 23.10.

This type of extinguisher is suitable for fires of all kinds whereas a water emitting extinguisher should not be used near to mains electrical supplies as mentioned on p. 58. Be careful that the extinguisher does not blow burning material about and do not touch the spout of the extinguisher once it is used — it gets very cold. It is also noisy in operation. Remember too that a lot of carbon dioxide gas in a small room could cause suffocation.

If the fire is not stopped immediately, then send for the fire service. If the fire is beyond your control evacuate the building, if possible closing doors and windows behind you.

Often when escaping from a burning building it is necessary to crawl out because hot air and smoke rise away from the floor. One's face can be covered with a wet cloth to facilitate breathing.

A person whose clothes are on fire should be laid down and the flames can then be smothered.

Large, wet towel or fire blanket is laid over the fire

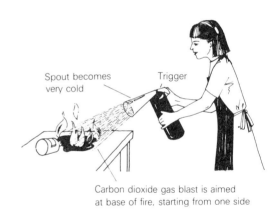

Spout becomes very cold

Trigger

Carbon dioxide gas blast is aimed at base of fire, starting from one side

Fig. 23.10 Putting out a small fire (a) using a fire blanket; (b) using a carbon dioxide fire extinguisher

SUMMARY

- Approach an injured person quickly but without panic.

- Speak confidently and clearly: 'Lie still. Where are you hurt?'

- Remove any hazards. Control onlookers.

- Make a quick investigation, start life-saving treatment if needed, and send for help.

- Make a second, more thorough examination and give appropriate first aid treatment.

- Serious injuries can cause death by cessation of breathing, severe bleeding, damage to internal organs, poisoning, extensive wounds or burns, or shock.

- The patient should be reassured and made comfortable.

- An obstruction in the throat may be expelled by coughing or by slapping the back.

- Artificial ventilation is usually achieved by the mouth-to-mouth method.

- External chest compression requires skillful, regular pressing upon the breastbone.

- *Haemorrhage* means bleeding. Severe bleeding from a wound is reduced by a dressing pressing upon the wound.

- Injury shock is treated with rest, warmth and reassurance.

- Small wounds are dressed after cleansing the surrounding skin.

- Large, deep or gaping wounds are lightly dressed pending expert medical attention.

- Burns and scalds require immediate cooling with cold water followed by a dressing.

- Electrical burns are often more serious than the surface burn indicates.

- Nose-bleeding is treated by sitting the patient up with head forward and nostrils held closed for several minutes.

- Blisters can often just be covered with a dry dressing.

- Insect bites and stings can be treated with cold water and antihistamine cream (after any visible 'sting' has been removed).

- Foreign bodies in the eye may be washed out or lifted out with a moistened length of twisted cotton wool.

- A fracture can cause bleeding, pain, shock or injury to surrounding tissues. Movement must not be allowed to cause further injury or suffering.

- A *dislocation* is a bone 'out of joint'.

- A *sprain* is damage to ligaments at a joint.

- A *strain* is damage to a muscle or tendon.

- Muscle cramp is treated by extending the muscle, by massage and warming.

- Head injuries can cause concussion and compression. Infection of head wounds could be serious.

- The recovery position is advised for the unconscious patient.

- Fainting is a temporary loss of consciousness. Recovery is quick. It can often be avoided by lowering the head or lying down.

- An asthma attack affects the breathing.

- Heart attacks are due to obstruction of the blood supply to the heart muscles. Pain in the chest may be a symptom.

- A stroke is caused by interruption of the blood supply in the brain. Some paralysis may result. Lay patient down. Ensure that she can breathe.

- During an epileptic fit the patient should be protected from injuring herself.

- Electric shock requires that electricity be switched off or patient be moved from it, using insulating material. Injuries are then treated.

- Sugar is needed to prevent diabetic coma.

- Drinks of water are given if someone has drunk a corrosive poison.

- Small fires can be extinguished using a fire blanket or a carbon dioxide fire extinguisher. Water extinguishers are not suitable for electrical fires.

- A person whose clothes are on fire should be laid down and the flames smothered.

EXERCISE 23

1. When examining an unconscious patient what are the most important things to look for?

2. What information should be given when calling for help by telephone?

3. Name three ways in which injuries can cause death.

4. Give two reasons why breathing may stop.

5. What are the main points to remember when giving artificial respiration by the mouth-to-mouth method?

6. What does heart massage involve?

7. What immediate action should be taken when serious bleeding occurs, e.g. from an arm?

8. Give four signs and symptoms of shock.

9. How should a small cut on a finger be treated?

10. What are the aims of first aid treatment for burns and what action should be taken?

11. How is nose-bleeding treated?

12. What usually causes a blister on the foot?

13. Distinguish between a dislocation, a strain and a sprain.

14. Distinguish between concussion and compression.

15. Suggest how an unconscious person may be positioned while awaiting an ambulance.

16. What is the cause and what is the treatment for fainting?

17. (a) What is the cause of a stroke?

(b) Suggest two signs or symptoms which distinguish a stroke from a heart attack.

18. What is the main aim of first aid when a person suffers an epileptic fit?

19. (a) What is the cause of a diabetic coma?

(b) How might you check that the patient is not simply drunk?

20. What needs to be done when a person suffers a severe electric shock?

21. What precautions should be taken when using a carbon dioxide fire extinguisher?

22. How would you deal with a person whose clothes are on fire?

APPENDIX

TABLE 1   SOME PREFIXES FOR SUBDIVISIONS AND MULTIPLES OF UNITS

mega (M)	meaning	1 000 000
kilo (k)		1000
centi (c)		1/100
milli (m)		1/1000
micro (μ)		1/1 000 000
nano (n)		1/1 000 000 000

TABLE 2   UNITS AND APPROXIMATE RELATIONSHIPS BETWEEN UNITS

16 dram = 1 ounce (oz)   =   28 gram
16 oz   = 1 pound (lb)   = 454 gram
14 lb   = 1 stone (st)   = 6.35 kilogram
 8 st   = 1 hundredweight (cwt) = 50.8 kilogram

1 litre (l)   = 1000 centimetre^3 (cm^3)
1 millilitre (ml)   = 1 cm^3
1 cubic foot (ft^3) = 6.2 gallon   = 0.0283 metre^3
1 gallon   = 8 pints   = 4546 cm^3
1 pint   = 20 fluid ounces (fl oz) = 568 cm^3
1 fluid ounce   = 480 minim   = 30 cm^3
1 minim   = 60 millimetre^3 (mm^3)
1 teaspoonful   = 4 cm^3
1 tablespoonful = 16 cm^3

1 inch   = 2.54 cm
1 foot   = 12 inches = 30 cm
1 mile   = 1.6 kilometre (km)

1 calorie   = 4.2 joule (J)
1 Calorie   = 4200 J
1 Btu   = 1055 J
1 therm   = 100 000 Btu
1 electricity 'unit' = 3.6 megajoule (MJ)
  (kilowatt-hour)

TABLE 3 LIST OF CHEMICAL ELEMENTS AND PROTON NUMBERS
(AS FAR AS LEAD)

1	Hydrogen (H)	42	Molybdenum (Mo)
2	Helium (He)	43	Technetium (Tc)
3	Lithium (Li)	44	Ruthenium (Ru)
4	Beryllium (Be)	45	Rhodium (Rh)
5	Boron (B)	46	Palladium (Pd)
6	Carbon (C)	47	Silver (Ag)
7	Nitrogen (N)	48	Cadmium (Cd)
8	Oxygen (O)	49	Indium (In)
9	Fluorine (F)	50	Tin (Sn)
10	Neon (Ne)	51	Antimony (Sb)
11	Sodium (Na)	52	Tellurium (Te)
12	Magnesium (Mg)	53	Iodine (I)
13	Aluminium (Al)	54	Xenon (Xe)
14	Silicon (Si)	55	Caesium (Cs)
15	Phosphorus (P)	56	Barium (Ba)
16	Sulphur (S)	57	Lanthanum (La)
17	Chlorine (Cl)	58	Cerium (Ce)
18	Argon (Ar)	59	Praseodymium (Pr)
19	Potassium (K)	60	Neodymium (Nd)
20	Calcium (Ca)	61	Promethium (Pm)
21	Scandium (Sc)	62	Samarium (Sm)
22	Titanium (Ti)	63	Europium (Eu)
23	Vanadium (V)	64	Gadolinium (Gd)
24	Chromium (Cr)	65	Terbium (Tb)
25	Manganese (Mn)	66	Dysprosium (Dy)
26	Iron (Fe)	67	Holmium (Ho)
27	Cobalt (Co)	68	Erbium (Er)
28	Nickel (Ni)	69	Thulium (Tm)
29	Copper (Cu)	70	Ytterbium (Yb)
30	Zinc (Zn)	71	Lutetium (Lu)
31	Gallium (Ga)	72	Hafnium (Hf)
32	Germanium (Ge)	73	Tantalum (Ta)
33	Arsenic (As)	74	Tungsten (W)
34	Selenium (Se)	75	Rhenium (Re)
35	Bromine (Br)	76	Osmium (Os)
36	Krypton (Kr)	77	Iridium (Ir)
37	Rubidium (Rb)	78	Platinum (Pt)
38	Strontium (Sr)	79	Gold (Au)
39	Yttrium (Y)	80	Mercury (Hg)
40	Zirconium (Zr)	81	Thallium (Tl)
41	Niobium (Nb)	82	Lead (Pb)

TABLE 4   SOME EXAMPLES OF SYSTEMATIC NAMES OF CHEMICALS

Common name	Molecular formula	Systematic name	Comments
ORGANIC CHEMICALS			
Acetic acid	CH_3CO_2H	Ethanoic acid	'oic' denotes the COOH or CO_2H group
Acetone	CH_3COCH_3	Propanone	
Benzene	C_6H_6	Benzene	
Di-ethyl ether	$(C_2H_5)_2O$	Ethoxyethane	
Ethyl acetate	$CH_3CO_2C_2H_5$	Ethyl ethanoate	
Ethyl alcohol	C_2H_5OH	Ethanol	
Ethylene	C_2H_4	Ethene	
Formaldehyde	CH_2O	Methanal	
Glycerol (or glycerine)	$C_3H_5(OH)_3$	Propan-1,2,3-triol	Also called propanetriol
Iso-propyl alcohol	$(CH_3)_2CHOH$	Propan-2-ol	
Methyl alcohol	CH_3OH	Methanol	
Phenol	C_6H_5OH	Phenol	
Stearic acid	$C_{17}H_{35}CO_2H$	Octadecanoic acid	'Octadecan' denotes 18 carbon atoms
Urea	$CO(NH_2)_2$	Urea or carbamide	
INORGANIC CHEMICALS			
Aluminium chloride	Al_2Cl_6	Aluminium chloride	
Borax (sodium borate)	$Na_2B_4O_7$	Disodium (I) tetraborate (III)	(I) and (III) denote number of electrons shared by sodium and boron atoms respectively
Carbon dioxide	CO_2	Carbon dioxide	
Hydrochloric acid	HCl	Hydrochloric acid	
Hypochlorous acid	$HClO$	Chloric (I) acid	
Hydrogen peroxide	H_2O_2	Hydrogen peroxide	
Ozone	O_3	Ozone or trioxygen	
Sodium carbonate	Na_2CO_3	Sodium carbonate	
Sodium bicarbonate	$NaHCO_3$	Sodium hydrogen carbonate	
Titanium dioxide	TiO_2	Titanium (IV) oxide	(IV) indicates the number of electrons the Ti atom shares with the other atoms

ANSWERS TO EXERCISES

Exercise 1

1. 5 kg, 4.4 lb **2.** (a) 90 (b) 0.9 (c) 3
3. (a) 0.0635 (b) 25.4
4. (a) 2000 (b) 0.002 **5.** 28 m²
6. 102-76-104 **7.** Inch, kilogram, pound, ounce
8. (a) 105 (b) 6 (c) 20 **9.** 3
10. 9.1 **11.** (a) 6750 (b) 6.75
12. 500 **13.** 7.5 kg **14.** 200
15. 44 **16.** 0.8% (or 700/16)
17. 14
19. (a) (i) 6 (ii) 60 (b) 60 N upwards (c) 60 N (d) 90 J
20. (a) 30 J (b) 70% **21.** 8 640 000
22. 3771 J/min **23.** 60 watt, 300 watt

Exercise 2

1. 5000 **2.** None **3.** (c)
4. Barometer **5.** Pressure difference
6. (a) Close stopcock (b) Stopcock
(c) Stopcock. Basin plug out.
7. All **8.** Neck, armpits or groin
9. (a) 6 kgf/cm² or 600 kPa (b) 6 kgf (60 N)
10. (a) Heart (b) Muscles (c) Muscles (d) Gravity

Exercise 3

1. (a) Electron (b) Oxygen molecule
2. Carbon, glass, polythene **3.** Hydrogen
4. Nucleus **5.** (a) H (b) Cl (c) Na
6. Hydrogen, oxygen, iron
7. Air, face powder, lipstick **8.** 5 + 3 = 8
9. (a) Mascara (b) Electrodes (c) Make-up
10. Page 21 **11.** 0.95 g/cm³, 0.95, float
12. Page 21 **13.** 111 cm³
14. Glycerine is denser than water; liquid paraffin is less dense

Exercise 4

1. Current **2.** Electrons **3.** Both
4. Spread out **5.** (a) — (b) +

6. Repulsion, attraction **7.** Equal
8. —, +, towards, bigger, attracted **9.** All
10. Shock, fire, clinging
11. (a) A (b) V (c) W **12.** Page 29
13. (a) 4.5 V (b) 1.5 V **14.** Zero
15. 20 Ω **16.** 24 V **17.** 6 V, 0.6 A
18. 4 Ω **19.** Muscle **20.** Page 27
21. 1/6 A
22. (a) 8 A, 1920 W, (b) 4 A, 960 W, (c) 2 A, 480 W
23. Pages 35 and 36 **24.** Page 36
25. Epilation, iontophoresis, disencrustation
26. (a) Page 36, 37 (b) Page 37
27. Protects instrument
28. Page 37
29. Epilation, high frequency skin treatment

Exercise 5

1. (a) Page 45 (b) Page 44 (c) Page 44, 47
(d) Page 44, 19
2. (a) Electrons (b) Na⁺ to cathode, Cl⁻ to anode
3. Sodium hydroxide **4.** Electrolysis
5. Page 46 **6.** Small area of needle
7. Softening, contact, absorption **8.** Negative
9. Page 46 **10.** Page 48
11. Chemical and heating **12.** Page 45

Exercise 6

1. DC, AC. Low volts, high volts **2.** L, N, E
3. L, N **4.** 50 alternations per second
5. 2.4 A **6.** Possible danger **7.** Shock, fire
8. Page 53 **9.** Safety
10. (a) Parallel (b) None (c) 12 A
11. Protects wiring and sockets **12.** Page 55
13. (a) Brown (b) Yellow and green **14.** b and c
15. Hands dry. Hold plug **16.** Page 59
17. Over-loading socket
18. Mains is sinusoidal **19.** Page 55
20. E conductor to N. Not clamped. Loose strand. Wire not stripped
21. Page 55 **22.** (a) DC (b) AC (sinusoidal)
23. (a) Electrical energy (Electricity bills) (b) 'unit'
24. £1.20

Exercise 7

1. (a) 100 °C   (b) 0 °C   (c) −5 °C   (d) 22 °C
2. 113 °F   **3.** 50 °C   **4.** 90 °C
5. 140 °F, 158 °F   **6.** Cracking glassware
7. Page 65, 66, 68   **8.** Low temperatures
9. Constriction, range, size
10. Page 65   **11.** Page 66   **12.** Page 66, 67
13. Page 67   **14.** Steam operates bimetal switch
15. Page 69   **16.** Joule, calorie, Btu (page 71)
17. Page 70   **18.** Page 71   **19.** Page 71
20.

100 000

10 000   1000

21. 10 260 000 Btu
22. (a) 720   (b) 270, 0.14 p, 0.54 p   **23.** Page 73
24. Page 70
25. (a) See question 3   (b) See question 2
(c) Page 63 and see question 4

Exercise 8

1. Page 78, 79   **2.** Page 81   **3.** Implosion
4. Page 82   **5.** Ventilation   **6.** Page 79
7. Page 12   **8.** In place of cup   **9.** Smaller reading
10. 2720 Pa   **11.** 420 cm   **12.** It collapses!

Exercise 9

1. Page 85, 88   **2.** Page 85   **3.** Page 86
4. Page 86   **5.** Expansion when water freezes
6. Close stoptaps, open sink taps, etc.   **7.** Page 88
8. Cooling by evaporation   **9.** Page 89
10. Page 91   **11.** Page 92   **12.** Page 93
13. Page 93   **14.** Stabilises emulsion (page 96)
15. Page 95   **16.** Page 95   **17.** Page 96
18. Page 97   **19.** Page 99   **20.** Page 99
21. Page 100   **22.** Page 100
23. (a) Page 21   (b) Page 100   **24.** Page 99
25. (a) 21 g   (b) 2.1 kg   (c) (i) 11.5   (ii) 13°   (d) 67%   (e) 25%

Exercise 10

1. Page 104   **2.** Page 105   **3.** Page 105
4. Mainly (a) Radiation   (b) Radiation and natural convection
(c) Forced convection
5. Page 111   **6.** Page 114   **7.** Page 113
8. Convection

Exercise 11

1. Page 119   **2.** 0.34 m   **3.** 40 dbA
4. Page 122   **5.** Page 122
6. Absorption, reflection
7. High pitch, high frequency, small wavelength

Exercise 12

1. Iron   **2.** Page 126   **3.** Page 128
4. Page 128   **5.** Page 131   **6.** Page 132
7. Page 134

Exercise 13

1. AC to DC   **2.** Page 146   **3.** Page 143
4. Page 137, 140   **5.** Page 140, 141   **6.** Page 143
7. Smoothing (page 142). Also see page 150
8. Resonance and spontaneous oscillation
9. Page 143, 150

Exercise 14

1. Page 154   **2.** Page 159   **3.** Page 155
4. Page 156   **5.** Page 158
6. (a) Page 158, 159   (b) Page 159   **7.** Page 159
8. Page 158   **9.** Page 160   **10.** Page 162
11. Epilation   **12.** Fire (page 163)   **13.** Page 163
14. Developing, printing   **15.** Page 164

Exercise 15

1. B, G, Y, O, R   **2.** 5×10^{14} Hz   **3.** Page 169, 170
4. 400 and 700 nm   **5.** R, G, B   **6.** R, Y, B
7. Y and B   **8.** (a) Black   (b) Black   (c) Yellow

Exercise 16

1. 400 lux   **2.** Epilation   **3.** Page 178
4. Page 178   **5.** Cheap to run; long life
6. Page 179

Exercise 17

1. IR and UV
2. Page 181 (a) All colours + UV   (b) No blue, green UV
(c) IR only
3. Radiant R + IR, true IR only
4. Penetrating, irritant
5. UVA IT and some DT. UVB DT   **6.** A and B
7. Page 187, 188   **8.** Page 185, 187   **9.** Page 183, 189
10. UV (keratitis)   **11.** Burns. Perspiration

Exercise 18

1. Page 193   **2.** Page 194
3. (a) Na, Cl   (b) Na, H, C, O   (c) Na, C, O
4. Page 196   **5.** Page 196
6. (a) Burns   (b) Fire   (c) Burns   **7.** Hydrochloric
8. Page 200   **9.** Ethyl alcohol   **10.** Ethanol
11. Page 199   **12.** Fragrance   **13.** Page 201
14. Sodium carbonate, "permutit". Page 201
15. Page 195, 202

Exercise 19

1. Carbohydrate, proteins, fats **2.** Page 216
3. Page 206 **4.** (a) Saliva (b) Page 210
5. Healthy blood vessels
6. 2700 Cal/day approximately (page 220)
7. (a) Carbon, hydrogen, oxygen (b) Sugar
8. Page 217 **9.** See crossword solution below

10. Bile production. Glycogen, production and storage
11. Excretion of water and waste chemicals

Exercise 20

1. Page 226 **2.** Page 229 **3.** Page 238
4. Oxygen, carbon dioxide, nutrients, defence, cooling/warming, hormones
5. Page 228 **6.** Page 231 **7.** Page 229
8. Page 231 **9.** Page 233
10. (a) Page 236 (b) Page 236
11. Lactic acid, muscle fatigue **12.** Page 238
13. Page 241 **14.** Page 245 **15.** Page 234
16. (a) Skull (b) Knee or elbow **17.** Page 246
18. See page 240

Exercise 21

1. (a) Mineral oil (b) Ethanol (c) Glycerine (d) Talc
(e) Talc
2. Sheep **3.** Page 251
4. Castor oil. Beeswax **5.** Page 43, 257
6. Nitrocellulose. Resin (page 257) **7.** Page 258
8. Page 261 **9.** Page 200, 272 **10.** Ethanol
11. Page 259, 260 **12.** Page 251

Exercise 22

1. Page 264 **2.** Page 265 **3.** Page 266
4. Influenza. Measles. Chicken pox. Common cold
5. Page 268 **6.** Page 269
7. Alcohol. Cetrimide, hydrogen peroxide **8.** Page 273
9. Selenium sulphide, zinc pyrithione
10. (a) Aluminium chlorhydrate, aluminium chloride
(b) Hexachlorophene
11. Page 274 **12.** Sterilisation
13. Sink. Trap (page 275) **14.** Page 271
15. Food poisoning **16.** Athlete's foot, ringworm

Exercise 23

1. Breathing, bleeding, broken bones
2. Help needed, location, nature of injuries
3. Page 279 **4.** Electric shock, suffocation
5. Clear mouth, tilt head, close nostrils, timing, ensure chest moves
6. Regular pressing on breast bone
7. Lay patient down, firmly applied pad
8. Feels faint and sick. Pale. Skin cold and clammy. Abnormal breathing and pulse
9. Clean around it. Dry dressing
10. Cool it. Dress it. Treat shock
11. Sit quietly. Hold nostrils **12.** Rubbing
13. Page 285 **14.** Page 286
15. Recovery position
16. Blood supply to brain. See page 286
17. (a) Bleeding in brain
(b) Paralysis in stroke. Chest pain in heart attack
18. Avoid injury
19. (a) Too much insulin/too little sugar
(b) Diabetic card. Sugar in pocket/bag
20. Stop electricity. Treat injuries (page 288)
21. Spout cold. Blows paper etc. Suffocation
22. Lay her down. Smother flames

INDEX